IMAGES
of America

THE CAREY SALT MINE

On the cover: Please see page 80. (Courtesy of George E. Roberts.)

IMAGES
of America

THE CAREY SALT MINE

Barbara C. Ulrich

ARCADIA
PUBLISHING

ISBN 978-0-7385-6180-6

Published by Arcadia Publishing
Charleston, South Carolina

Library of Congress Catalog Card Number: 2008929880

For all general information contact Arcadia Publishing at:
Telephone 843-853-2070
Fax 843-853-0044
E-mail sales@arcadiapublishing.com
For customer service and orders:
Toll-Free 1-888-313-2665

Visit us on the Internet at www.arcadiapublishing.com

To my family—those who came before me,
those who grew with me, and those who have followed

Contents

ACKNOWLEDGMENTS

The problem with history is that there are so many people involved in the process of passing down the collective memory that it is impossible to thank them all—or often to even know who they are. The bulk of the images in this book come from the archives of the Reno County Historical Society, which let me use them without restriction. But its archives was built from the contributions of hundreds of people who recognized the significance of the relics from their past and cared enough to pack them into the car, drive them to the museum, and offer them up to the public domain with no compensation and little recognition. Many thanks go to those people as well as the museum staff and volunteers who have spent countless hours documenting and preserving the artifacts. Unless otherwise noted, all images come from the Reno County Historical Society.

Thanks also go to Lee Spence and Chris Eden at Underground Vaults and Storage for so graciously sharing their time, knowledge, and institutional photographs. Mike Carey was also extremely generous in lending his family scrapbooks. And I am completely indebted to the Hutchinson Salt Company for allowing me to explore the mine and poke through the Carey Salt Company papers for days on end. Myron Marcotte, the mine's underground supervisor, deserves special recognition for his saintlike patience and constant willingness to answer even the dumbest of questions—usually ending each answer with the reassuring line, "I just made all of that up."

And finally, thanks to all those who, tired of my whining, cleared my schedule of kids and household and prodded me with such helpful telephone messages as "less hooky, more booky." You truly were an inspiration.

INTRODUCTION

In 1901, Emerson Carey installed two grainer pans at his ice and cold storage plant at Avenue C and Main Street in Hutchinson. He obtained a charter for his new corporation, the Carey Salt Company, for the "manufacture of ice and salt." His was the last salt company to be established in Hutchinson, 26 years after salt was discovered in Kansas. It was also one of the most enduring. The rock salt mine, the last of the Carey Salt Company, was finally sold and became the Hutchinson Salt Company in 1991, 90 years after its formation.

Carey owned several businesses throughout his lifetime, including a strawboard and egg case filler plant, a railroad, a bag company, an ice and cold storage company, a coal company, a newspaper, a printing company, a construction materials company, a salt company, and a cemetery. All those businesses were successful, but none was as well known as the Carey Salt Company. Like all the other salt companies, he first sank some brine wells into the salt bed and began his evaporating plant. By the 1920s, he had scores of salt products to sell but saw a real need for rock salt to supplement some of his other products, particularly in the agricultural market. In April 1922, the Carey Salt Company announced plans to dig a shaft nearly one mile east of his new state-of-the-art evaporating plant. A total of $300,000 of 7 percent preferred stock would be sold to finance the project. In two days, the issue was oversubscribed, and just over a year later, the mine opened amid much fanfare.

The Carey Salt Company came to include Carey's first plant at Avenue C and Main Street in Hutchinson (which quit producing salt in the late 1920s), the newer east evaporation plant, and the mine. In 1931, to tap into markets in the south, Carey opened a rock salt mine in Winnfield, Louisiana, and then in the 1960s the Cote Blanche evaporating plant and mines, also in Louisiana. By that decade, consumers could purchase Carey Salt in 27 states and into Latin America.

Howard J. Carey took over the salt company when Emerson Carey passed away in 1933. Under his administration, departments were formed and important upgrades made to the facilities to increase production. After he retired, it was his son Howard J. "Jake" Carey's turn. In 1969, the Carey family sold the Hutchinson facilities, which by this time included the evaporation plant and the mine and the Hutchinson and Northern Railway (H&N) to the Interpace Corporation. Over the next two decades, the mine and evaporation plant changed hands several times. Interpace sold to Process Minerals, which sold to the North American Company. In 1991, the Hutchinson Salt Company purchased the mine only, breaking apart the mine and evaporation plant for the first time. But in the minds of many, it remains Carey Salt.

The topic of Emerson Carey and his collection of businesses is vast and spans over a century. His salt company alone is too broad a subject to fit into a single book, and this book will discuss the operations at the Carey Salt Mine only. When put on paper, mining is a very simple process. Miners get the room ready to blow up, they blow it up, they clean up the mess, and they sell the mess. In the 80 years of the mine's history, these tasks never changed, although the equipment became so efficient over the years that now 5 men do what 18 used to do underground, with a much higher yield. Many of the photographs that describe this process are not in chronological order or are not dated. This is because the processes have been so long-lived that the dates are not as important.

The 1950s seem to have been a golden age for the Carey Salt Company. In this era, the company hired a staff photographer to document the events at the Hutchinson, Lyons, and Winnfield facilities. This remarkable collection of over 1,800 photographs and negatives, which are now in the hands of the Reno County Historical Society, comprise the bulk of the images in this book. The photographs span the 1950s, yet, since the mining process remained the same, they apply to the 1960s and 1970s as well.

One

Constructing the Mine

As early as 1912, people were whispering that Emerson Carey was thinking of building a rock salt mine to go with his shiny new evaporating plant, located just outside the city limits east of town. The second decade of the 20th century was a difficult time for the Carey Salt Company. Battles against salt trusts and difficulty in dealing with railroad freight rates were making it tough to turn a profit, and since the company had been chartered in 1901, Carey was forced to use profits from his ice and other businesses to absorb the losses from salt company.

By 1922, however, all that was behind it. A location nearly a mile east of the evaporation plant was chosen, and in June, Carey Salt broke ground and began to dig the shaft for its new mine.

Many in Hutchinson had contemplated a rock salt mine, but all feared the thick aquifer running between the earth's surface and the salt bed some 400 feet below. Carey Salt chose the Foundation Company from New York to dig the shaft because of its success in building skyscrapers on the watery bedrock of the Big Apple. Even the Foundation Company wanted an extra $30,000 for the job.

Howard J. Carey, Emerson's son, chose the Allen-Garcia Company, an engineering firm in Chicago, to design the mine, which consisted of the entire operations underground and a mill building that would process the salt and facilitate its distribution. Correspondence between Allen-Garcia and Howard J. Carey flew fast and thick between 1922 and 1923, when every detail from the hoist motor selection to the loading dock doors was contemplated.

Everything progressed smoothly until a month before the opening, when flu epidemics and manufacturing problems caused most of the equipment to be shipped late. Workers scrambled to finish before the governor came to town for the big dedication ceremony in July 1923. Howard J. Carey complained to Allen-Garcia about its poor attention to detail: "We are getting pretty much peeved about this job."

The company made it through the opening, though not much worked properly on that day. Allen-Garcia returned in coming weeks to work out the kinks, and the mine was open for business.

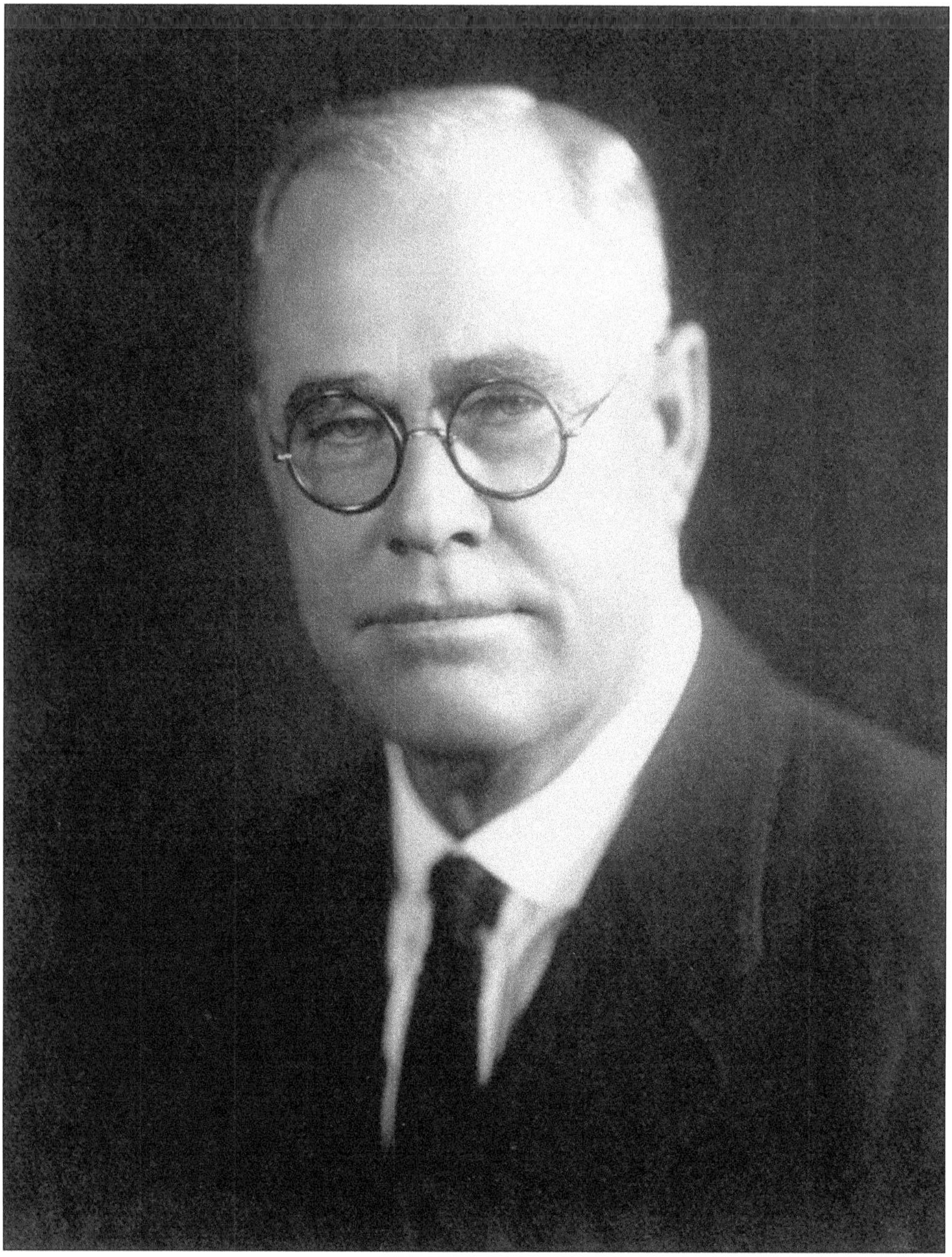

Emerson Carey could have come from a Horatio Alger novel, a character whose combination of pluck and luck brought him from his arrival in Hutchinson with a mere two bits in his pocket to his becoming a multimillionaire with several business empires and a career in state politics.

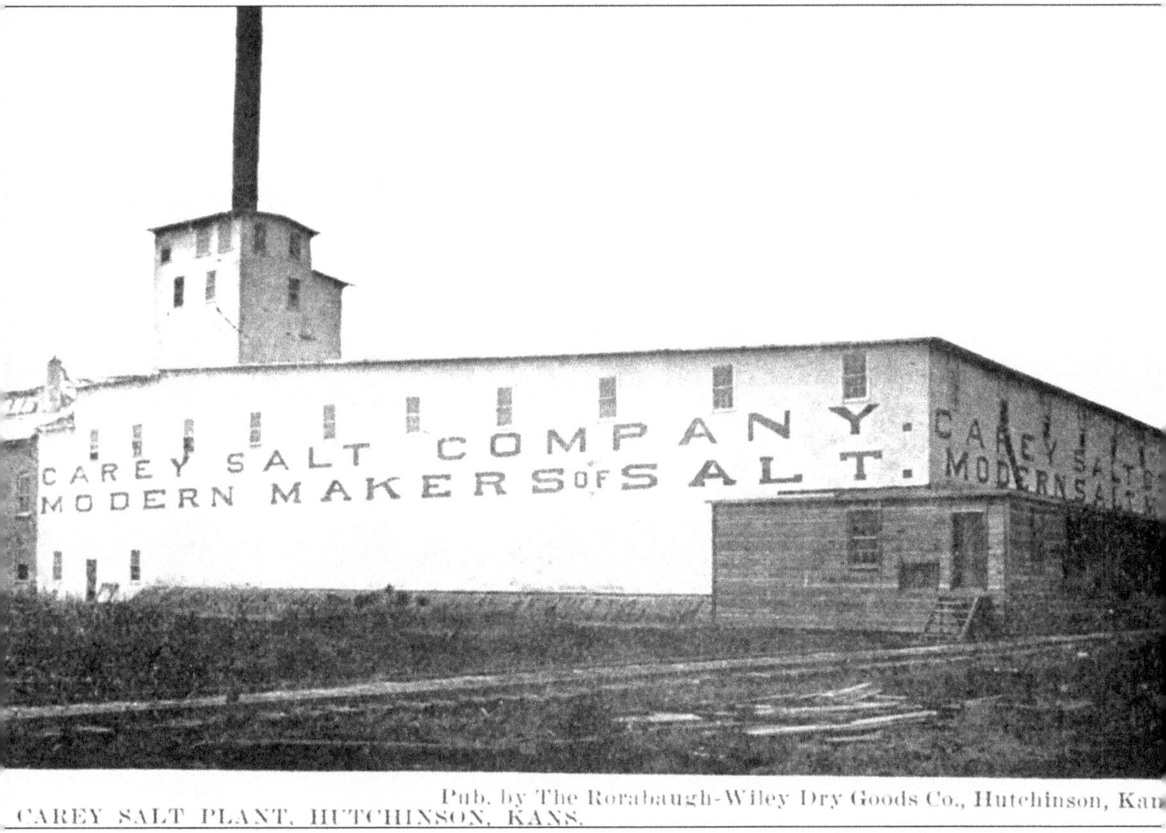

In 1901, Carey, who hated to waste anything, installed two grainer pans at his ice plant and used the steam from the ice process to evaporate the brine. The Carey Salt Company was chartered in April of that year "for the manufacture of ice and salt." Emerson Carey erected a new plant east of town in 1909, just outside the city limits. He installed state-of-the-art equipment and continued to expand. By 1912, rumors began to circulate that Carey Salt was looking for a location for a rock salt mine to go with the evaporation plant. This postcard shows the new east plant.

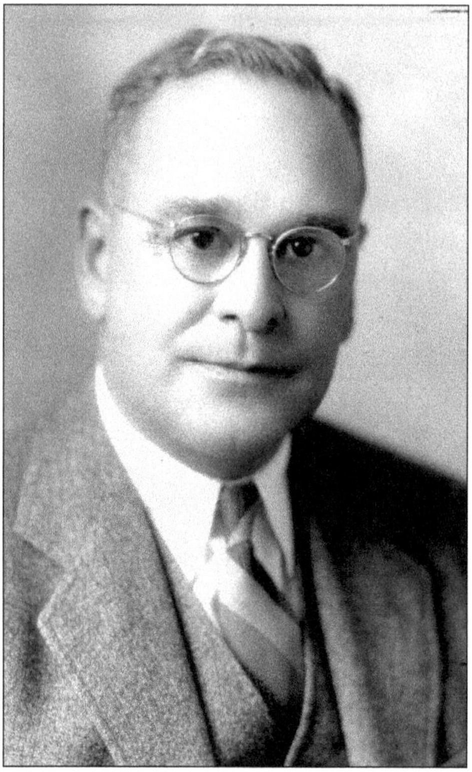

Emerson Carey spent over 20 years developing his salt company, but by 1922, he had at least in part handed the reins over to his son Howard J. Carey. The mine was Howard's project from the beginning. Even Howard's father deferred meeting with the engineers when Howard was unavailable.

The new mine would be located less than a mile from the evaporation plant, east of town. A single shaft would be dug nearly 650 feet below the earth's surface to bring rock blasted out of the vast Permian salt bed to the surface. Topside, a mill would grind and separate the salt into grades for sale to various industries. Howard J. Carey chose the Allen-Garcia Company, an engineering firm from Chicago, to design the facility.

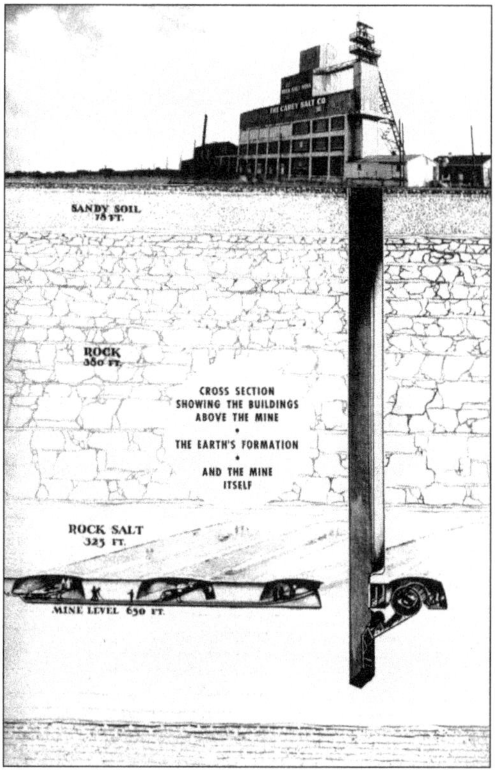

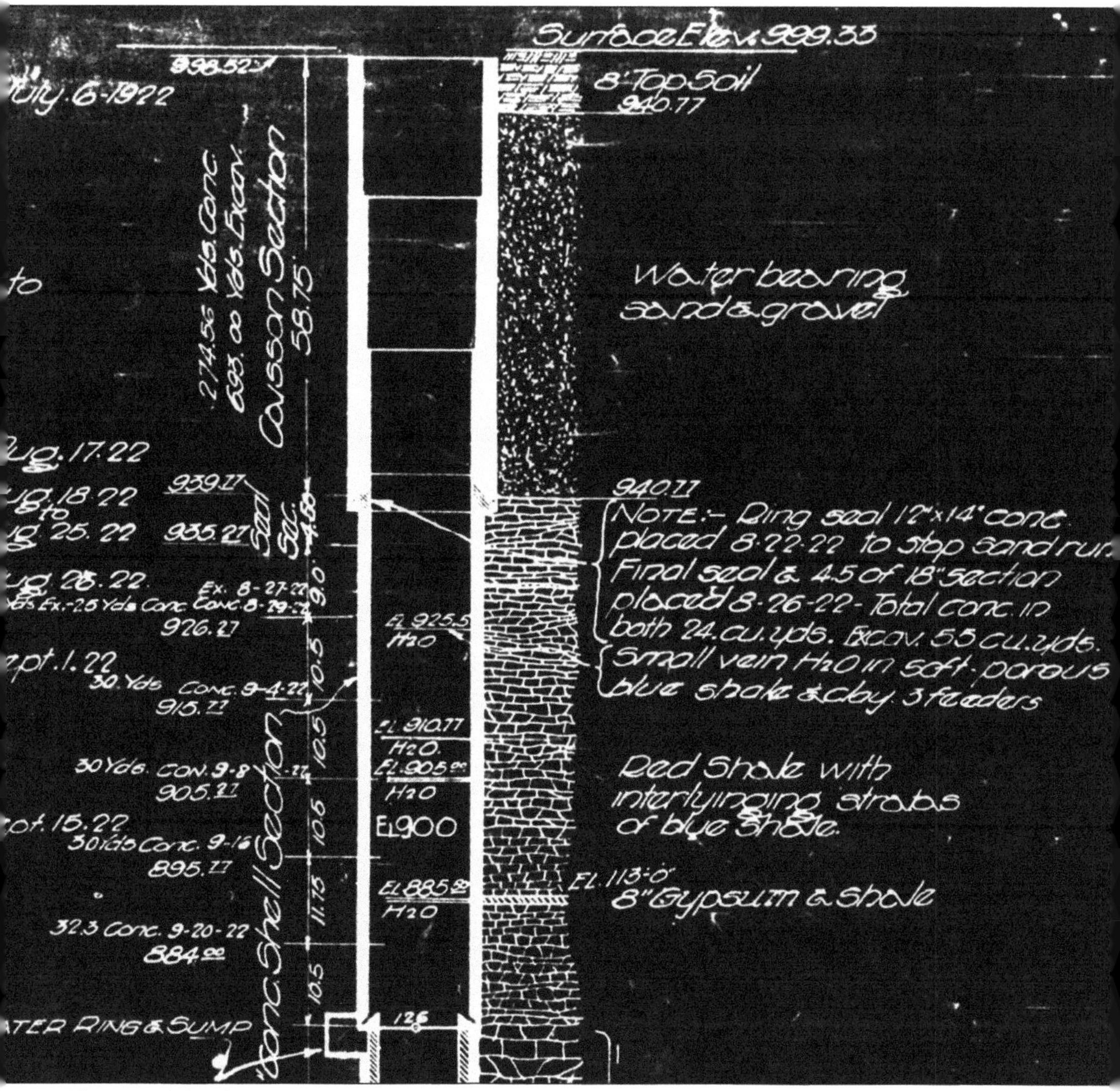

The top of the shaft was round, about 14 feet in diameter, and encased in concrete through the water table and several feet of rock beneath it. The Foundation Company employed a deep-sea diver to cut through the water table and lay the concrete. The bottom of the concrete casing had a ring to collect condensing or dripping water in case the concrete were to leak. The water drained into a sump that was then pumped to the surface to keep the shaft—and the salt below it—dry. The workers from the Foundation Company kept a log documenting their progress, eventually drafting a blueprint of their work. This image shows the top portion of that blueprint. The sump can be seen at the bottom of the picture.

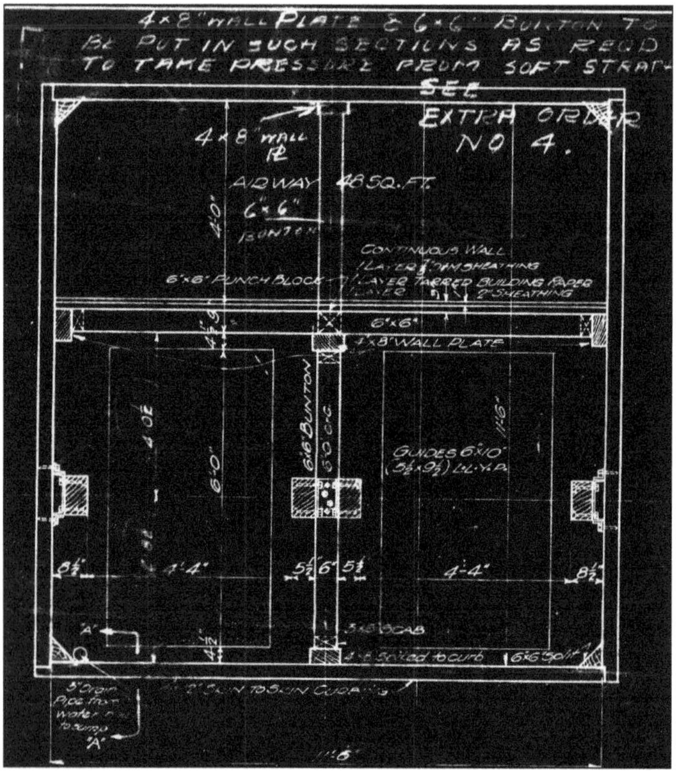

This blueprint shows the lower portion of the shaft, which was square and lined with timber. Air flowed into the mine through the back half of the shaft and exhausted through the front, which was divided for each skip, or bucket, of salt. The hoist would raise one bucket to the surface as the other lowered to mine level. Wooden guides would keep the skips from turning on their journey up the shaft.

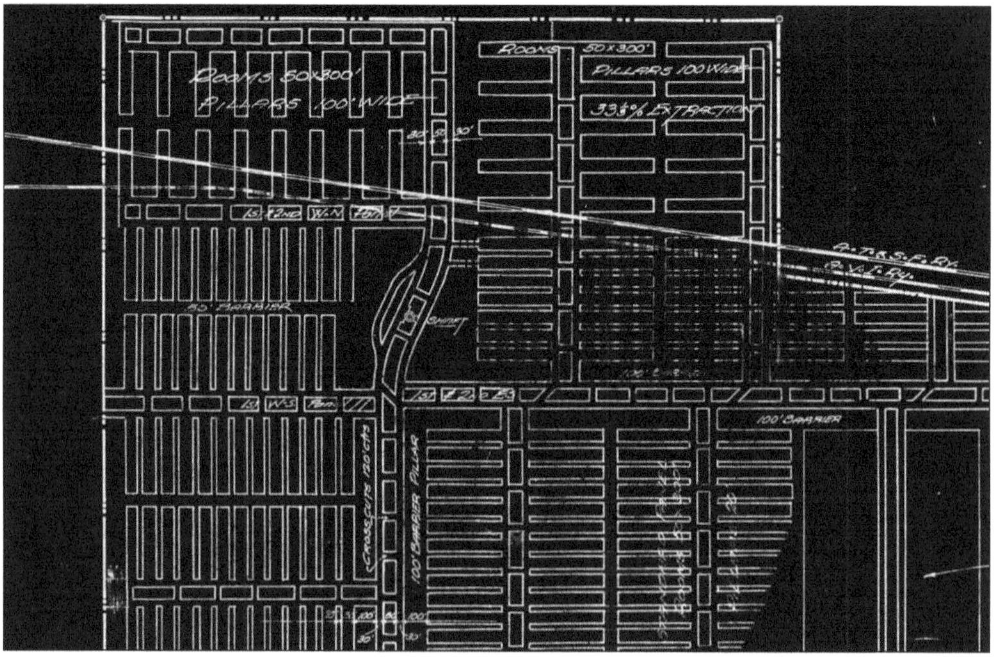

Underground, the mine's layout was as carefully considered as the mill aboveground. All the Allen-Garcia Company's recommendations left room for the mine to expand its equipment and its capacity for production. This mine development plan was drafted by Allen-Garcia to indicate the room layout.

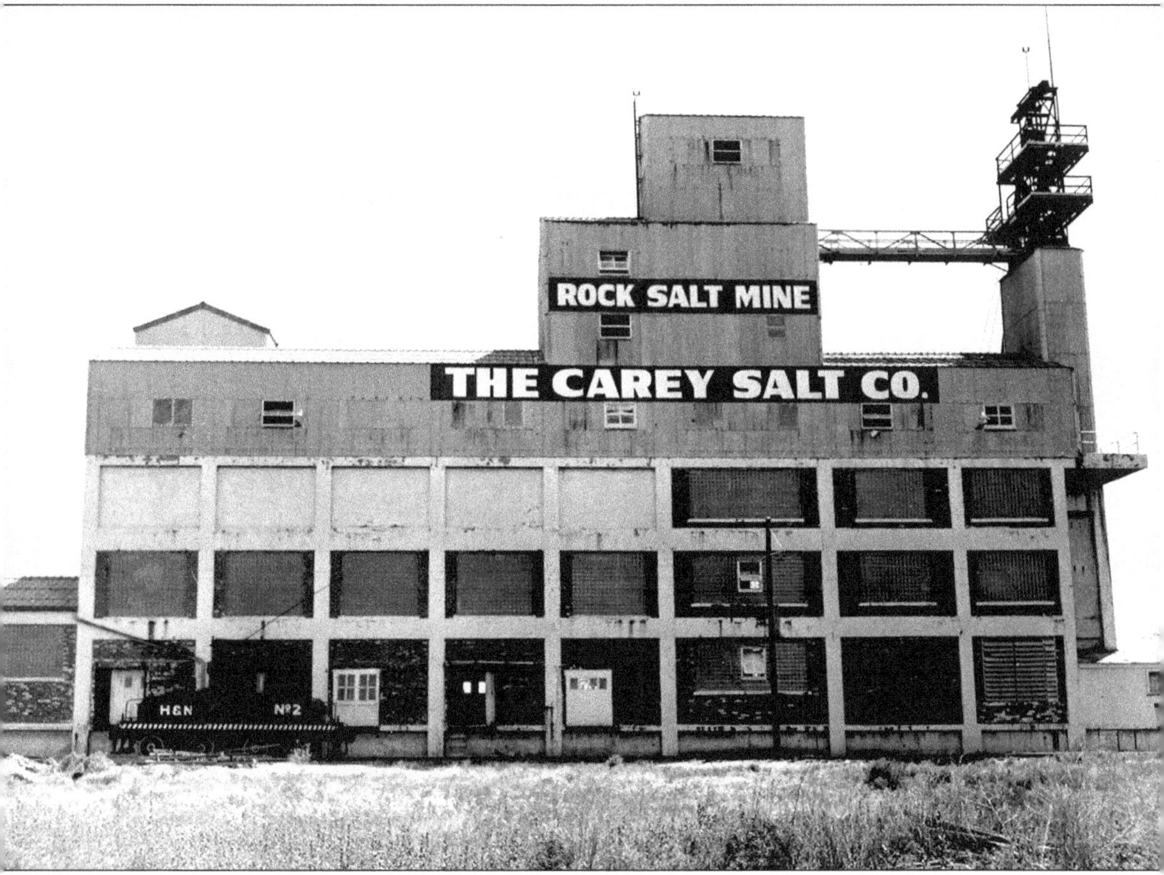

This view shows the original mill and tipple building. At the top right is the steel headframe, which sits directly over the shaft to hoist the skips. The skips dumped the salt at the top of the tipple building into a crusher, which led to an elevator located in the tall area of the mill building, center. The salt traveled down a series of chutes to crushers and screens, which separated the salt into sizes. The left side of the mill building contained the bins, which held the salt until it could be sold in bulk. The mill had docks for loading the railcars along both sides of the building.

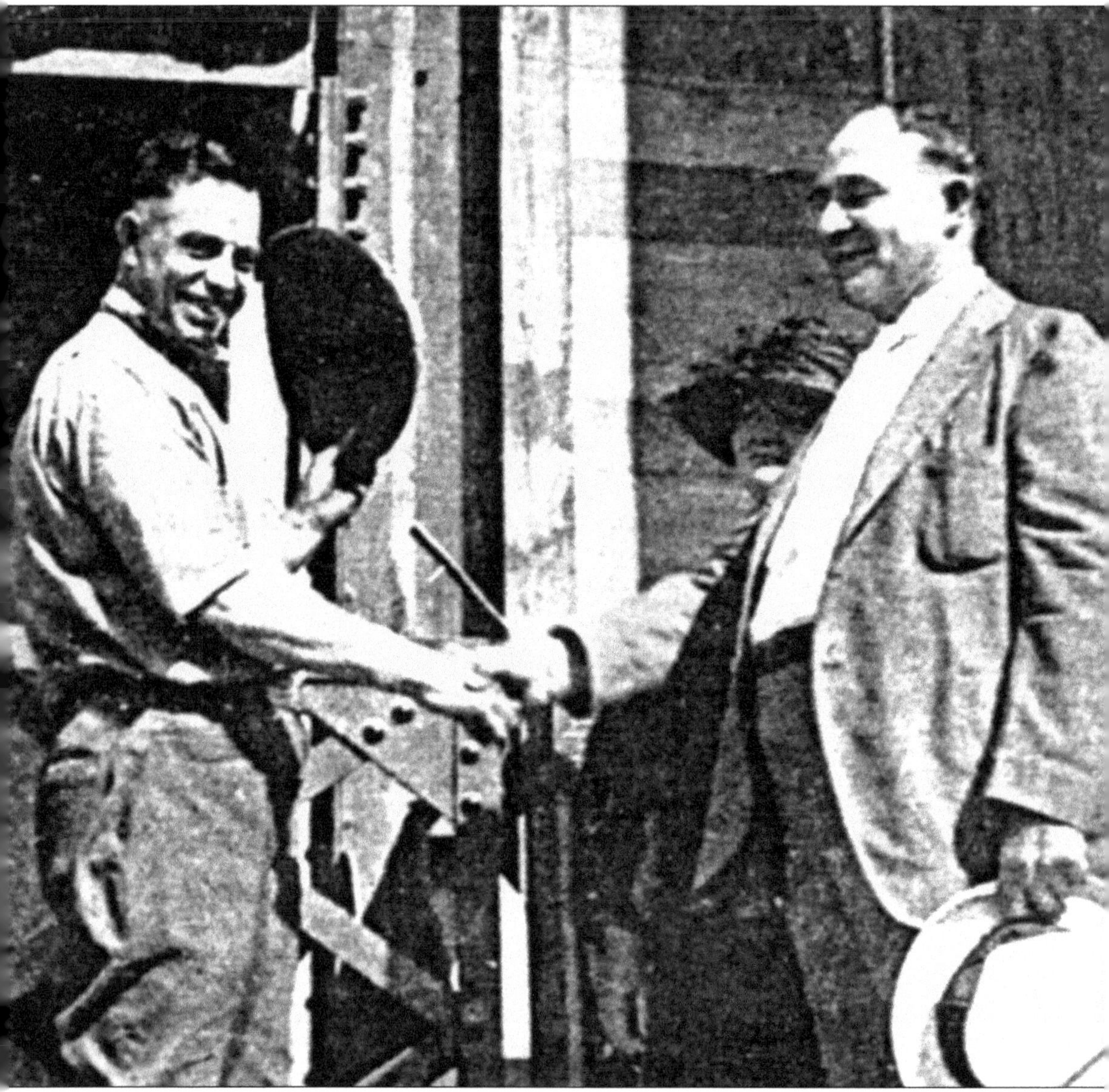

On June 23, 1923, Kansas governor Johnathan M. Davis pushed the button that raised the first load of salt at the mine's dedication ceremony. This photograph shows Charlie Walters (left), mine superintendent, shaking hands with the governor. The event was not without its blunders. Much of the equipment had been delayed in shipping, and miners were scrambling to finish installations by the 23rd. There was no time to work out the kinks. Later, a representative of the Allen-Garcia Company wrote to Howard J. Carey, "It was certainly a disappointment to me to have this crusher stalled on the day of your formal opening—especially since the Governor of your state happened by just as the belt began to slip and the crusher to clog."

Two

The Rail System and the Hoist

The hoist and underground railroad worked together to provide access from the face, to the area being mined, to the earth's surface. The salt came from the mine face by railroad and up the hoist to the mill to be processed. Going the other direction went supplies, equipment, and labor. The hoist and railroad were the lifeblood of the mine, and without either, the mine would not function.

The railroad was a small-gauge system. At first, the track crews laid the rails directly to the face within the bay so that loaders could fill the ore cars by hand. Later mechanized loading enabled the salt to be taken to the ore cars on the main lines, instead of having to lay the rails all the way to the face. This reduced the need for the track crews.

Three locomotives are known to have existed at the Carey Salt Company. The first ran solely on batteries. In 1936, during a series of upgrades, a larger engine was purchased that could pull more cars. After a particularly bad breakdown, it became apparent that Carey Salt needed a backup locomotive, so a third was purchased from a mine at Kanopolis. The track crews primarily used this engine.

In 1983, Carey Salt abandoned the cumbersome rails and switched to a conveyor belt system to carry the salt to the hoist. Diesel-powered vehicles were introduced to move the miners and supplies.

The Allen-Garcia Company purchased the hoist from the Nordberg Manufacturing Company. The hoist, which ran on a 200-horsepower motor, was located in a dedicated building aboveground and ran the two skips, or buckets, up and down the shaft. The Foundation Company ended the shaft about 37 feet below mine level to accommodate gravity-fed chutes that would dump the salt into the skips. The hoist lifted the skip to the tipple, located at the top of the adjoining mill building, to dump its four tons of salt to be processed into a marketable product. Changes in mining practices increased productivity, and Carey Salt sped up the hoist in the mid-1930s and again in the 1950s.

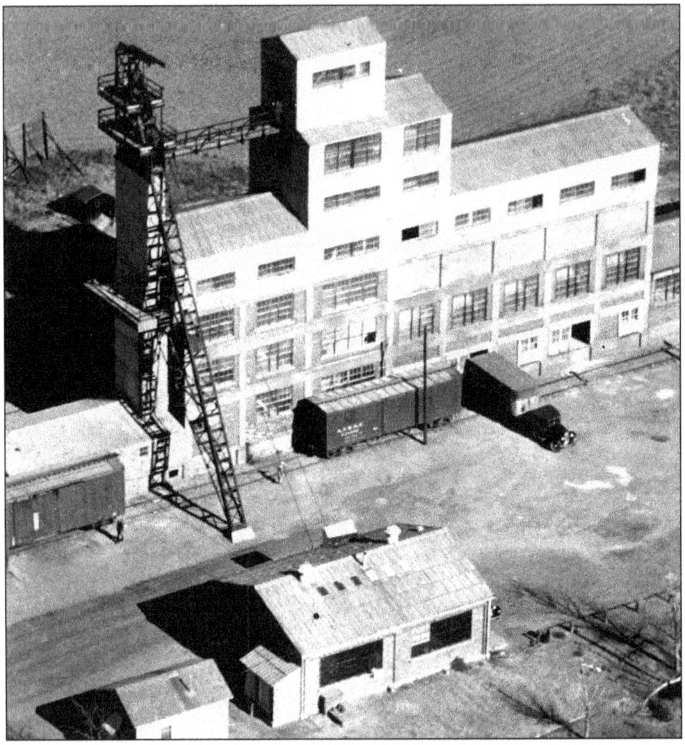

The hoist lifts and lowers the skips and the man-cage inside the shaft by winding and unwinding cables around a drum. In this pre-1944 photograph, the hoist house, where the hoist and drum are located, is in the foreground. The two cables, one for each skip, protrude from the building, stretching up to the metal headframe on the left side of the mill.

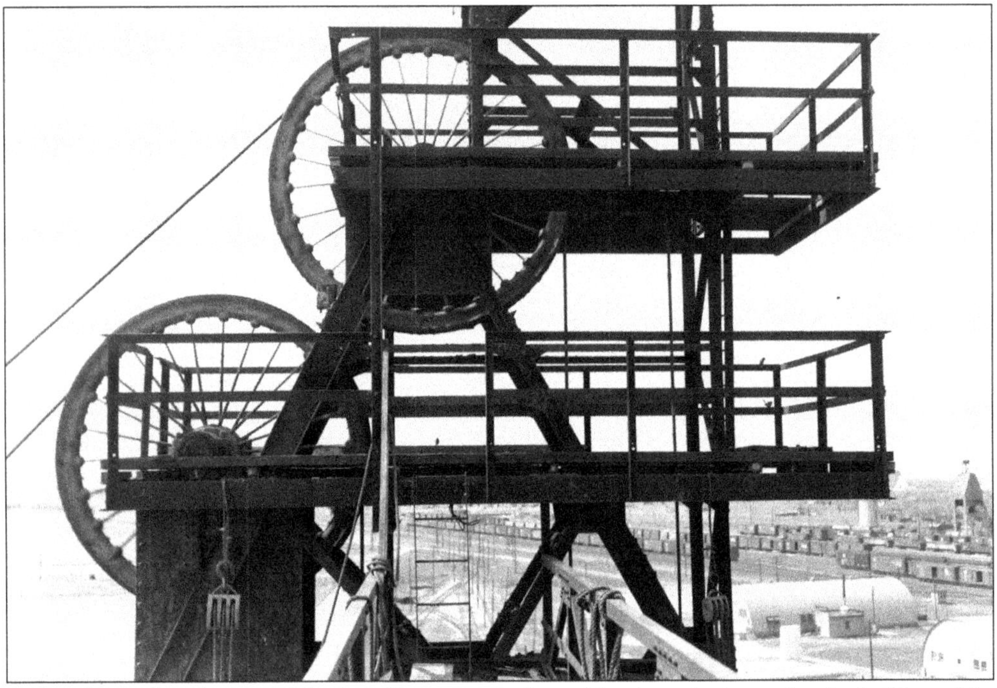

Pulleys located at the top of the headframe guide the cables directly down the shaft. The cable, which is called a rope by miners, is 1,000 feet long to allow for the depth of the mine shaft, the trip to the top of the mill where the skip dumps its salt, and the distance to the hoist house and around the drum.

Occasionally the pulleys must be replaced. This 1951 photograph shows a pulley being raised into place. The triangular metal frame was known as the back leg, which was part of the headframe. The Allen-Garcia Company specified that the headframe be made of medium steel and fastened with three-quarter-inch rivets.

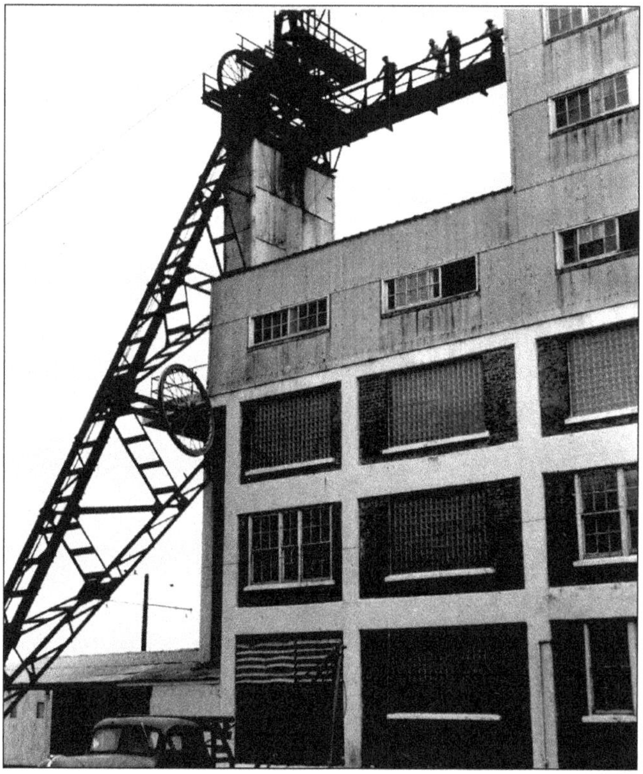

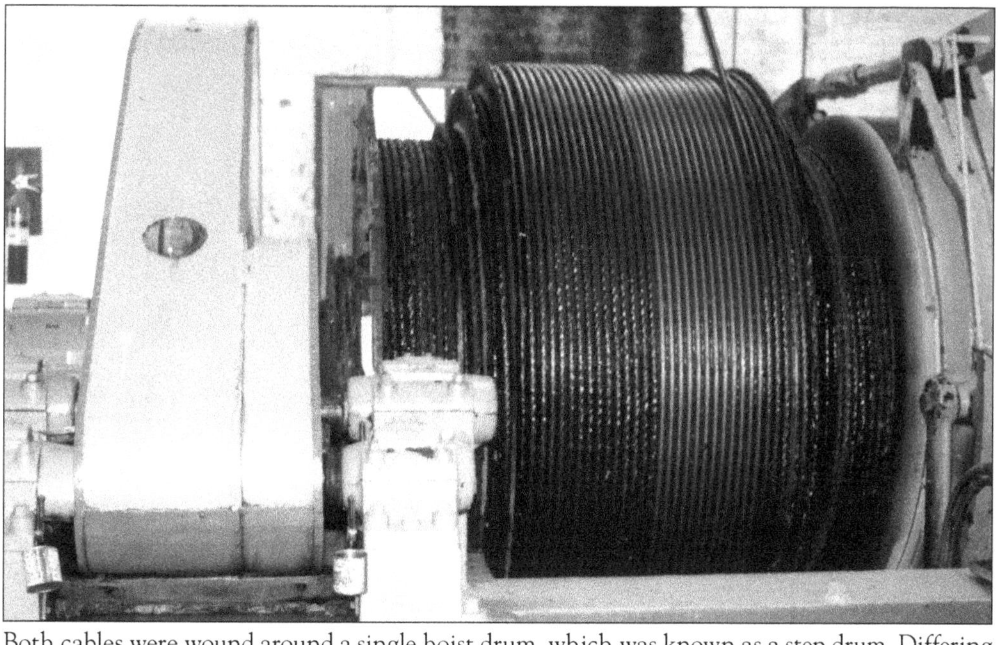

Both cables were wound around a single hoist drum, which was known as a step drum. Differing diameters along the drum itself made a smoother, more energy-efficient lift. As one skip rose to the surface, its cable wound on the drum. At the same time, the other skip was lowering underground, unwinding its cable along the drum. Brakes mounted on the drum held the skips in place.

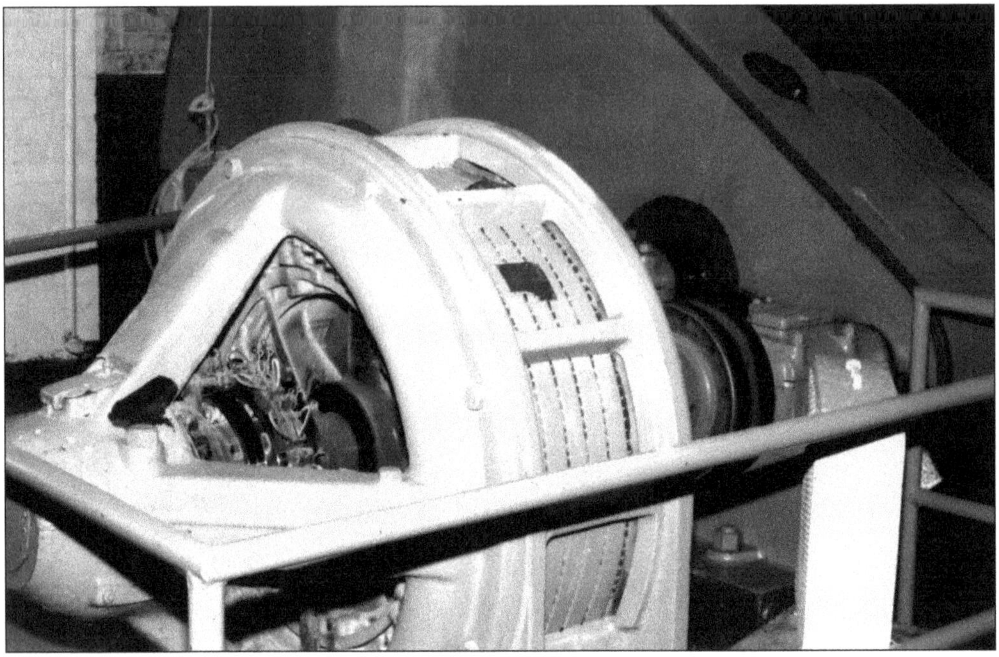

The Allen-Garcia Company specified that the hoist motor have 200 horsepower, even though in 1923 the Carey Salt Company did not need one that powerful. The mine was built for expansion, and in 1935 and again in the 1950s, the hoist was indeed sped up to keep up with increased production.

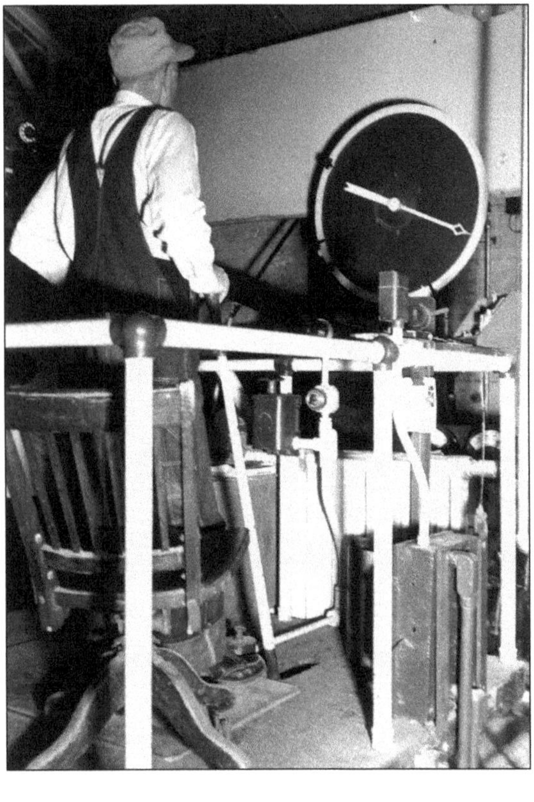

The hoist controls were operated manually. In this 1952 photograph, hoist operator Charlie Hungerford is running the hoist in the hoist house. The dial in front of him indicates the location of the skips within the shaft. At the time this picture was taken, the controls had just been replaced. For the first time, the hoist operator was able to sit down at his job.

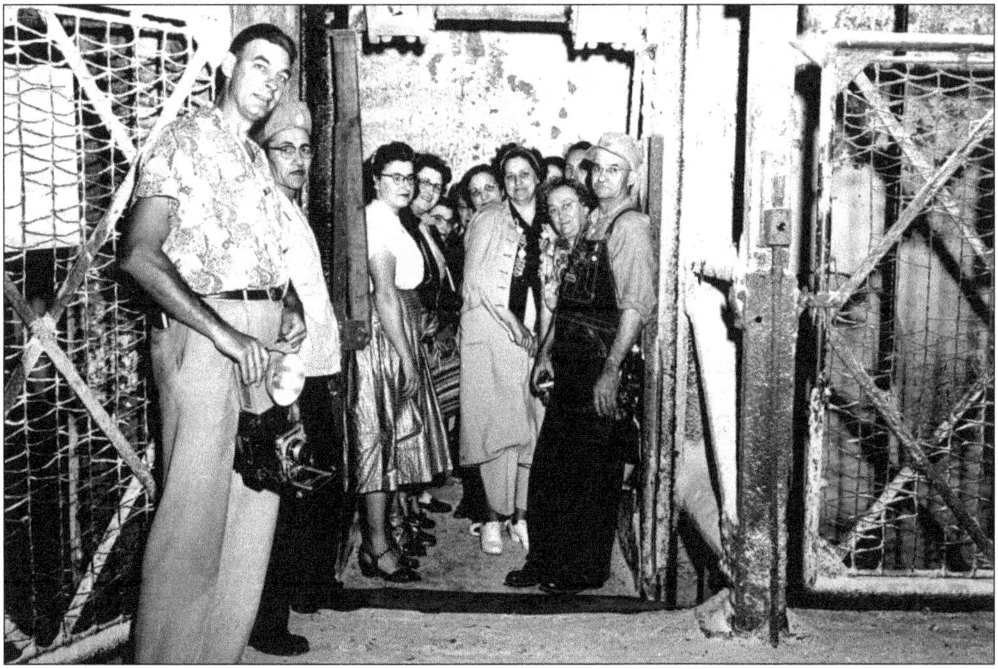

A signalman stayed at the shaft belowground. Through a system of bells, he communicated with the hoist operator when it was safe to raise the hoist. This 1952 photograph shows the signalman posing with a group of visitors. A sign with the bell signals is posted between the skips. The man-cages, which held about 10 people, hung beneath each bucket and date to at least 1927.

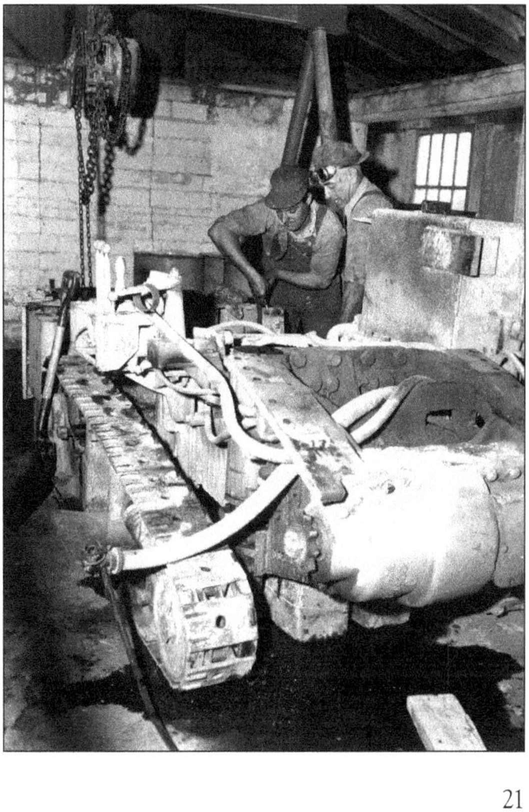

The shaft's partitions made it difficult to bring equipment underground. Available space totaled approximately a four-by-five-foot area. If equipment was too big to hang underneath the skip and man-cage, miners had to cut it into pieces small enough to fit. The Carey Salt Mine employed men that were able welders. This photograph shows Earl Bush (left) with an unidentified miner, disassembling a loader to fit down the shaft.

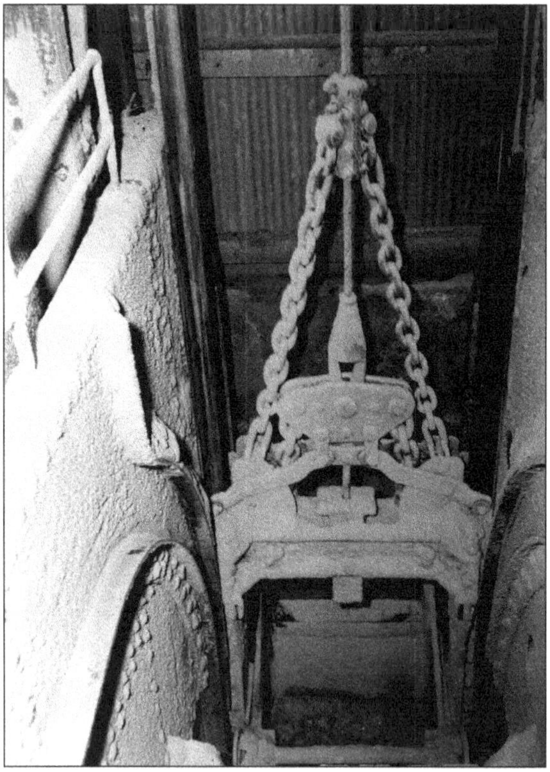

The skips were attached to a framework (shown here) called the bail, which held the skips and man-cage below it steady on its journey up and down the shaft. The top of the skip disengages from the bail as it glides forward on the tracks that dump the salt, which is called the tipple. On its way down, the skip slides back into the bail and reattaches itself with latches called dogs.

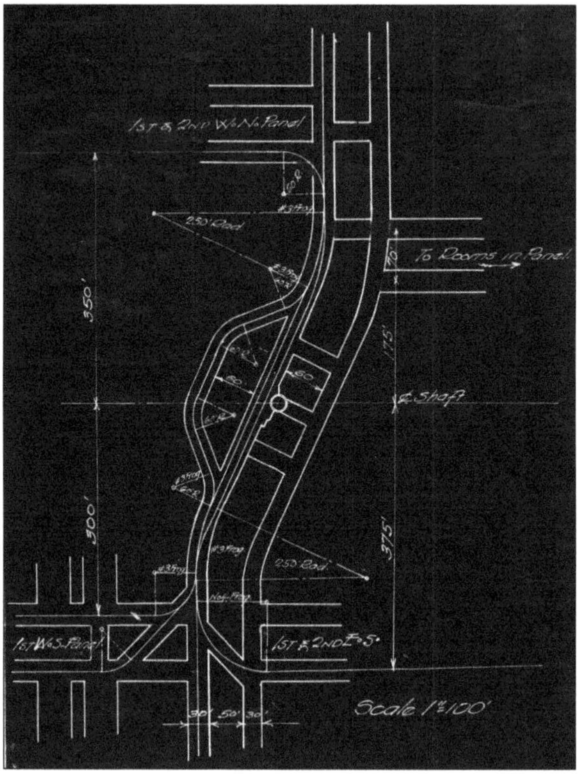

The Allen-Garcia Company designed the shaft area underground. In this 1922 blueprint, a circle represents the shaft itself. A double track ran next to the shaft for waiting cars, and the miners blasted a passageway for a track siding to allow empty trains to circumvent the full ones on their return trip to the face. The incomplete map shows the passageways to the bays that would be blasted in the first years of the mine.

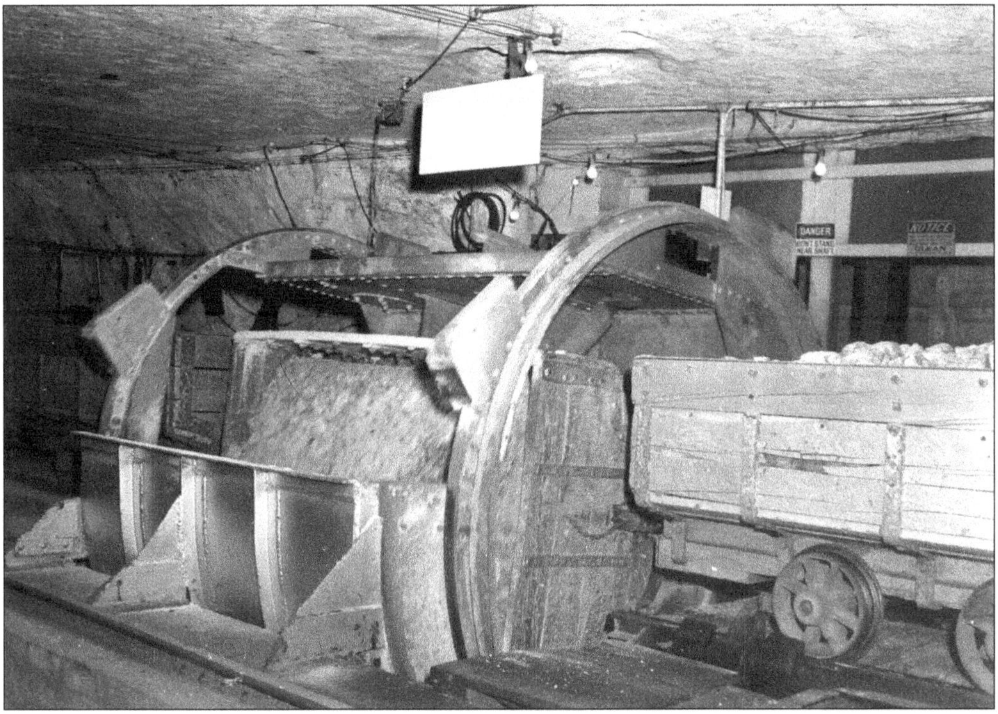

At the hoist, each ore car passed through the rotary dump that rotated it individually 135 degrees, emptying the salt into a chute that led to the skip.

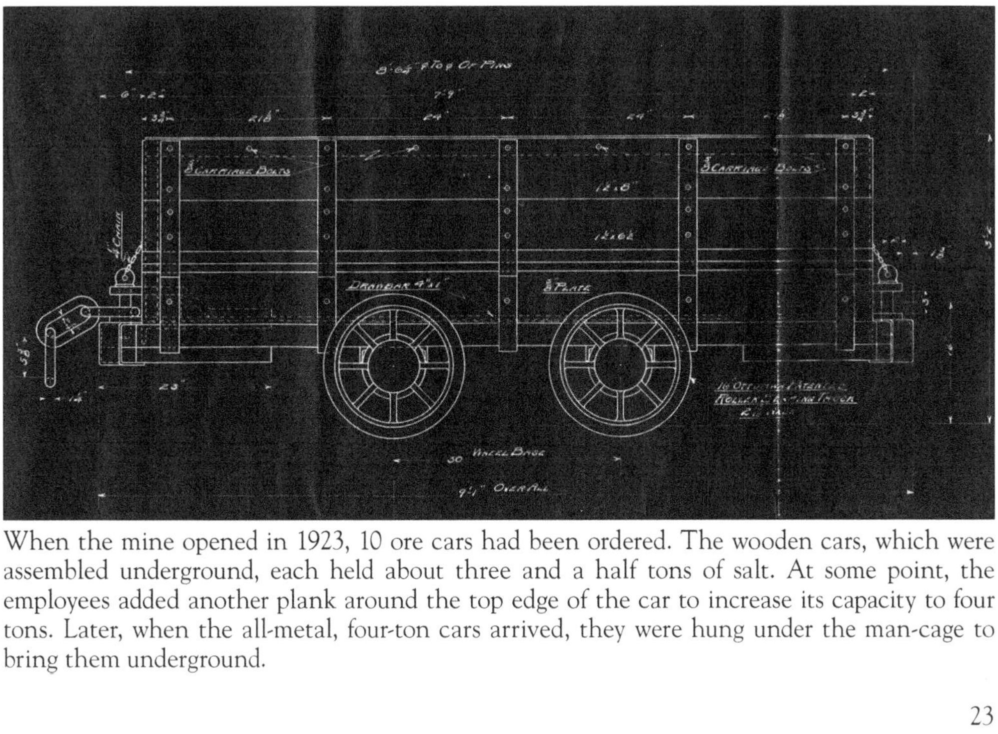

When the mine opened in 1923, 10 ore cars had been ordered. The wooden cars, which were assembled underground, each held about three and a half tons of salt. At some point, the employees added another plank around the top edge of the car to increase its capacity to four tons. Later, when the all-metal, four-ton cars arrived, they were hung under the man-cage to bring them underground.

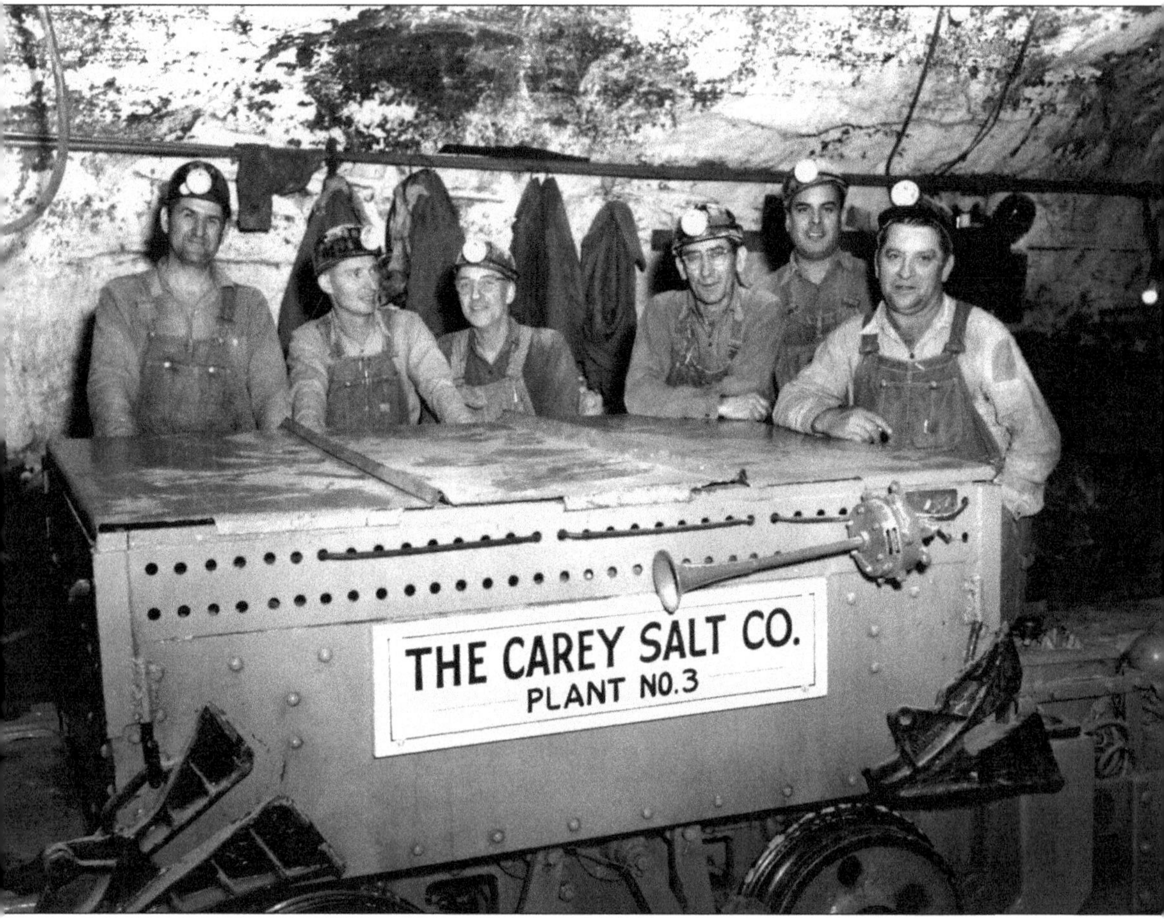

Evidently Howard J. Carey decided to wait until 1925 to purchase the Carey Salt Company's first locomotive. This General Electric engine ran on batteries and operated until the day the rails closed in 1983. Its power was limited, however, and when Carey Salt decided to mechanize, the engine could not pull the longer strings of cars. It then pulled the mantrip and other small cars. Standing in this 1956 photograph are, from left to right, Vernon Horton, Willie J. Thrift, William E. Gillock, John Thiessen, Bob Rodriguez, and Rudy Philbrick.

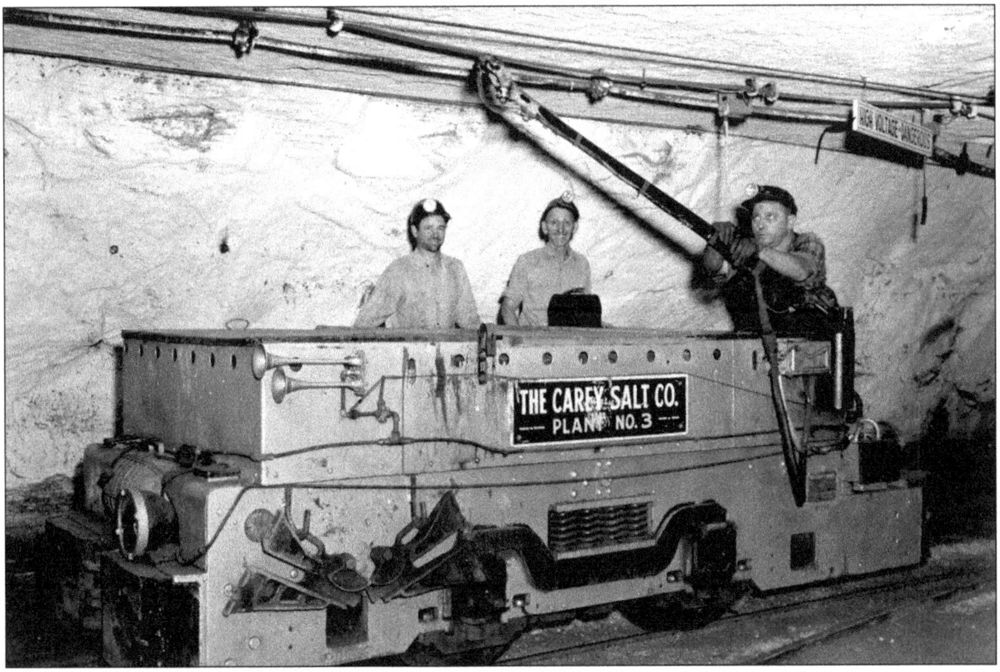

In 1935, Carey Salt purchased the main workhorse of the mine, this General Electric engine. It was battery operated and also ran on a direct current trolley line. In this photograph, engineer Rudy Philbrick is replacing the graphite lubricating shoe that makes contact with the trolley wire as Lee Boshart (left) and Fred L. Swepston look on.

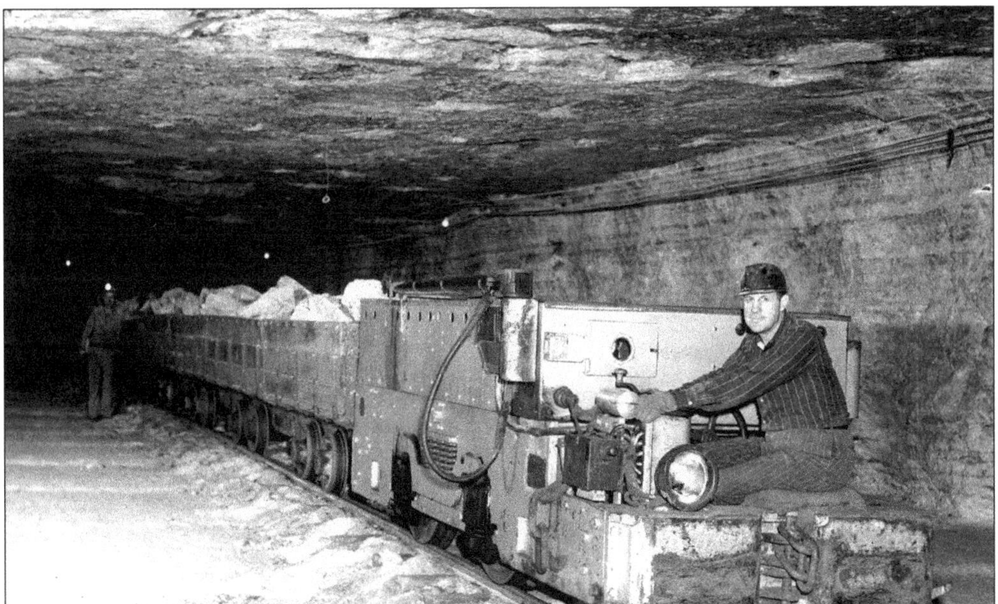

The new engine could carry a string of 13 full cars. The engine usually pulled the cars to the hoist and pushed them on their return trip to the face. In this 1940s photograph, Steven B. Horrell sits at the controls. The absence of trolley wire and the boom's lowered position indicates that the train was somewhere near the hoist when the photograph was taken.

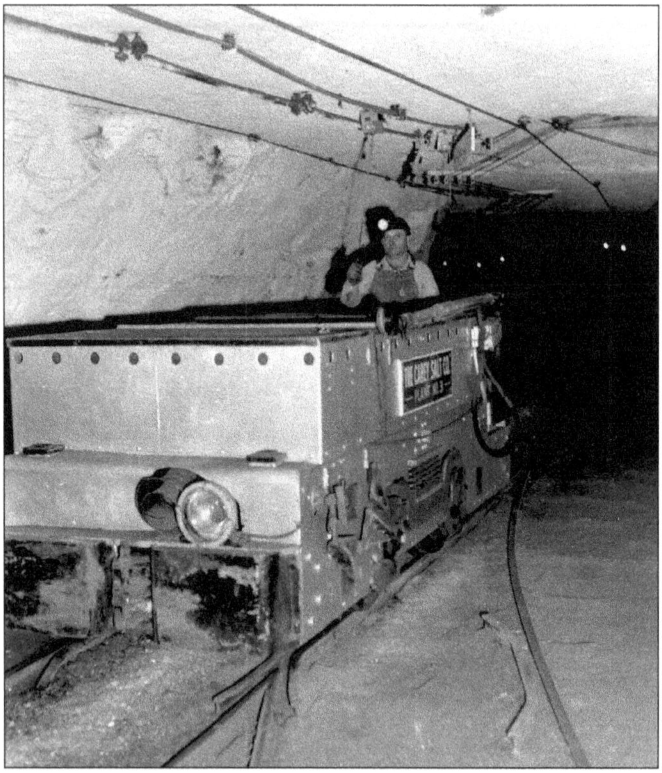

This photograph shows Rudy Philbrick pulling a switch to change tracks. This type of switch, called an electric switchman, did not require the operator to get off the train and was used at the hoist and at the car-loading station, the busiest places in the mine.

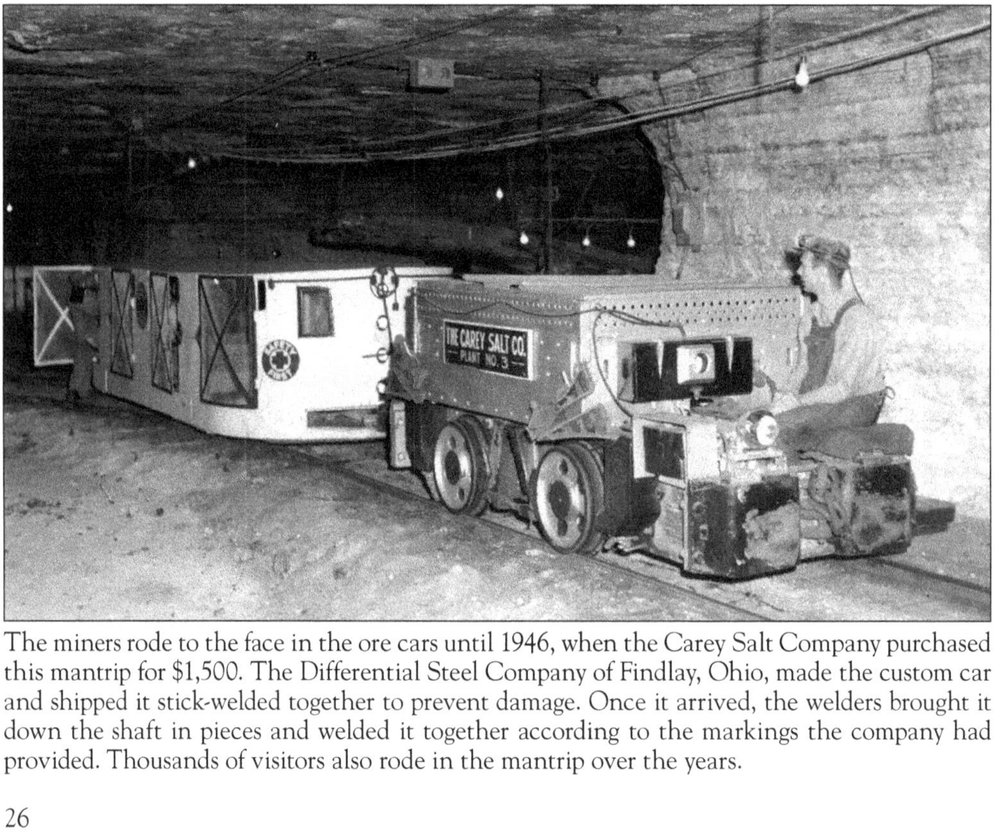

The miners rode to the face in the ore cars until 1946, when the Carey Salt Company purchased this mantrip for $1,500. The Differential Steel Company of Findlay, Ohio, made the custom car and shipped it stick-welded together to prevent damage. Once it arrived, the welders brought it down the shaft in pieces and welded it together according to the markings the company had provided. Thousands of visitors also rode in the mantrip over the years.

Three

MINING

The process of mining salt is simple—miners get the room ready, shoot it with explosives, and haul the loose salt to the surface for processing. But working underground is not the same as working aboveground. The absence of natural light and ready power underground make the task difficult, as do the ever-growing distances and limited access provided by the single, narrow shaft.

Over the decades, however, miners have used the same type of equipment to ready the room and to loosen the rock from the face, or the wall in which the workers are mining. The first step entailed using an undercutter to carve a slot in the face. The slot gave space for the rock to fall, making the blast more efficient. Miners drilled holes in a particular pattern along the face. The powderman then filled the holes with explosives, which blasted the salt into large chunks. The equipment used during the decades of the mine has become more powerful, efficient, and refined, but it is still the same basic process.

Hauling the salt away, on the other hand, has changed dramatically, from first using workers with picks to using diesel-powered vehicles with six-ton scoops. In 1941, the Carey Salt Company purchased an arm loader that swept the salt off the floor like a hungry insect, dumping it into a shuttle car that took it to the main rail line. No longer did the track crew have to lay track all the way to the face. The arm loader–shuttle car system reduced the need for hand loaders and for track layers at the same time the mine was harvesting salt like never before. Finally the conveyor belt line in 1983 marked the end of the railroad era and the accompanying arm loader and shuttle car. In 1923, the Allen-Garcia Company estimated that 18 men would be required to mine the salt underground. This did not include the blacksmith, handyman, trolleyman, hoist runner, or shift header. Today there are only five men who work at the face, pulling many times more salt out of the earth.

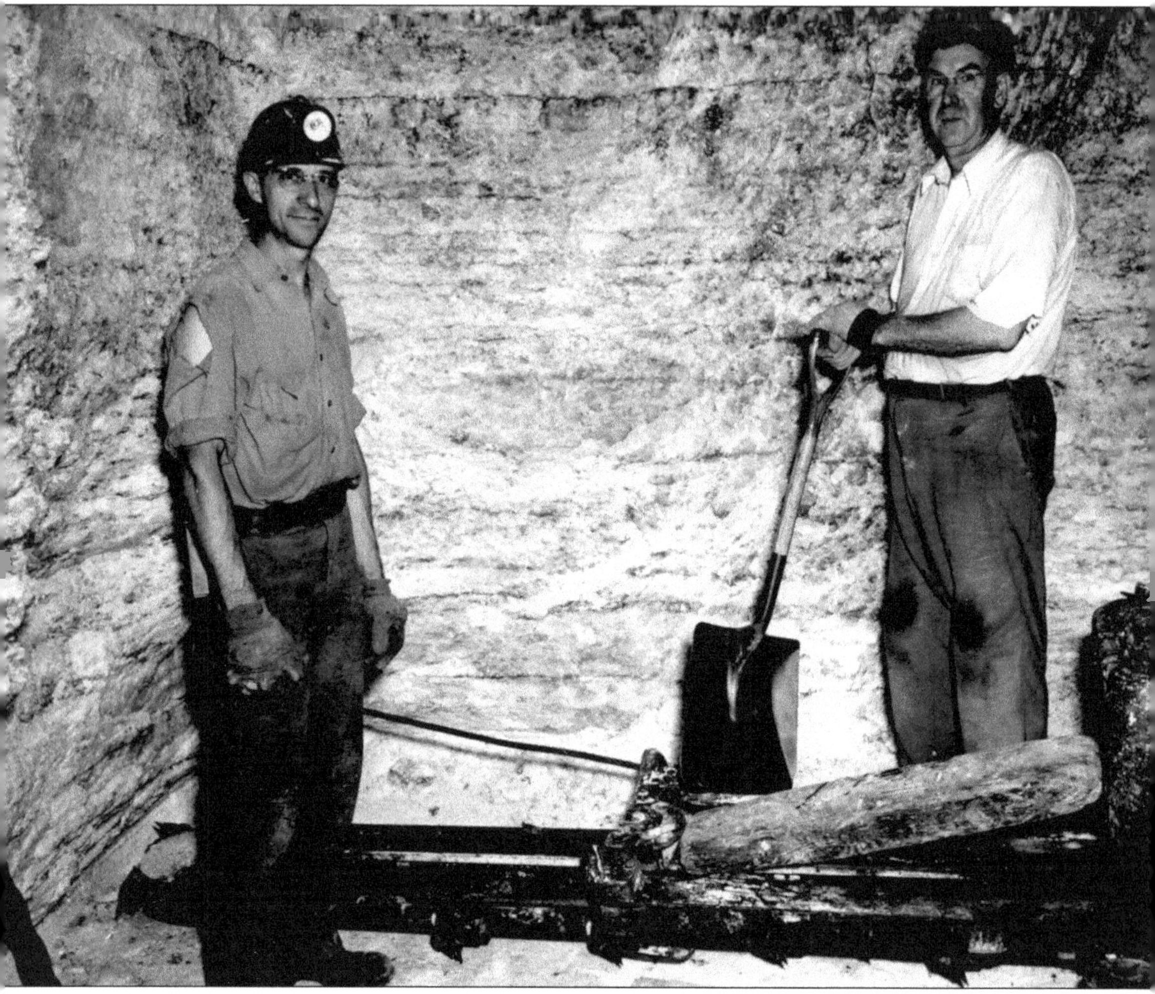

This World War II–era photograph shows Russ Wenger and Ike Brown, a salesman from the Carey Salt Company's Omaha district, posing by the undercutter. The picks on the cutting blade can clearly be seen. During the war, the salesmen participated in a campaign to work at the mine for two weeks. It is not clear whether this was done to provide relief from the overall labor shortage or if it was merely to give the sales force new insight into their products.

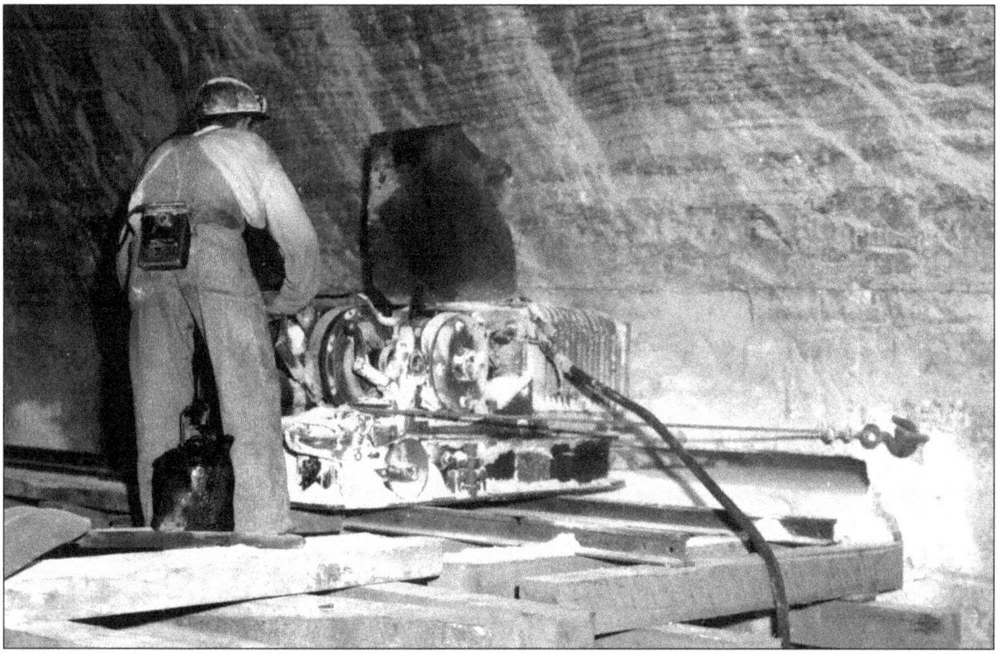

The undercutter was attached by a cable to a pin on each side of the proposed cut. Then one cable unwound from a reel as another wound, pulling the machine along at an even tension. Here a miner cuts midway up the wall to make a stage for the Atomic Energy Commission (AEC) experiments (see page 121). The scaffold is called cribbing and is used for a variety of purposes underground.

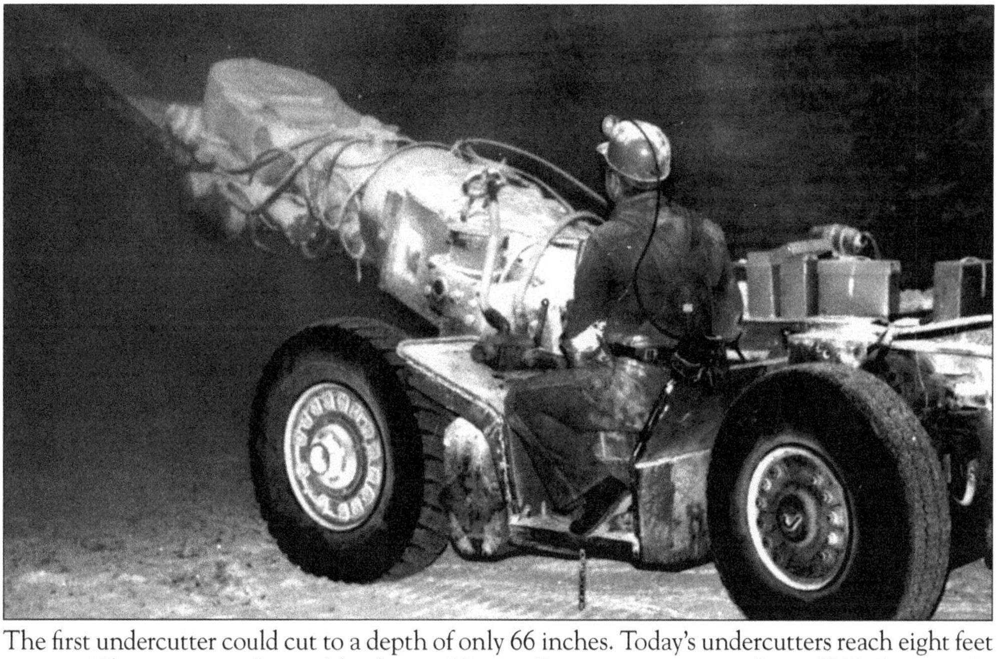

The first undercutter could cut to a depth of only 66 inches. Today's undercutters reach eight feet or more. The newest machines, like the one Harvey Coats is operating in this *c.* 1980 photograph, are hydraulic, can change direction and cut at different angles, and, although tethered to the power source by a cord, have much freer movement than the early undercutters had.

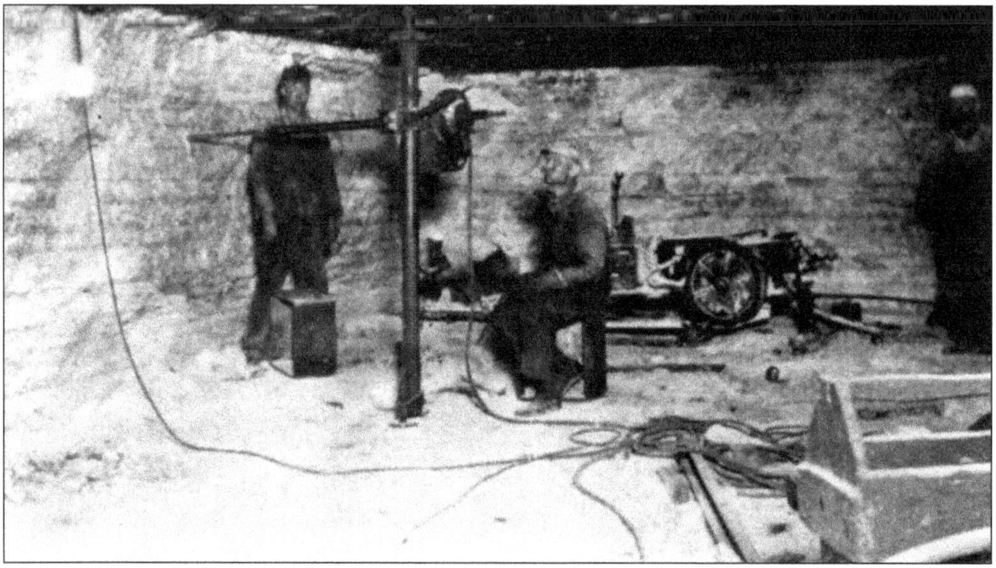

This photograph, which dates to about 1928, shows the earliest drill setup, where a post anchored to the floor and ceiling held a single electric drill attached to a large auger and a drill bit. One room could take hours to finish. The early undercutter sits in the background of this photograph.

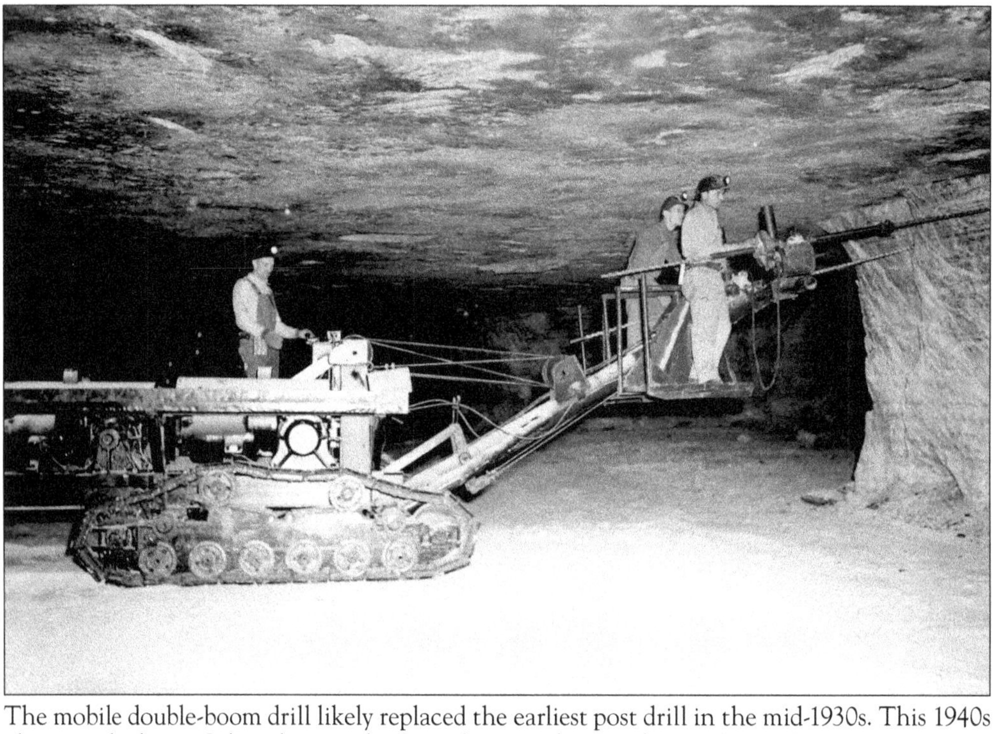

The mobile double-boom drill likely replaced the earliest post drill in the mid-1930s. This 1940s photograph shows Orlis Johnson, district salesman, driving the machine while Wallace Tracy (front) and Carroll Roxburg (back), Omaha district, operate the drills. The machine saved considerable time and reached the high targets more easily. Once this drill had been replaced by newer drills, the tractor crawler was transformed into a powder car.

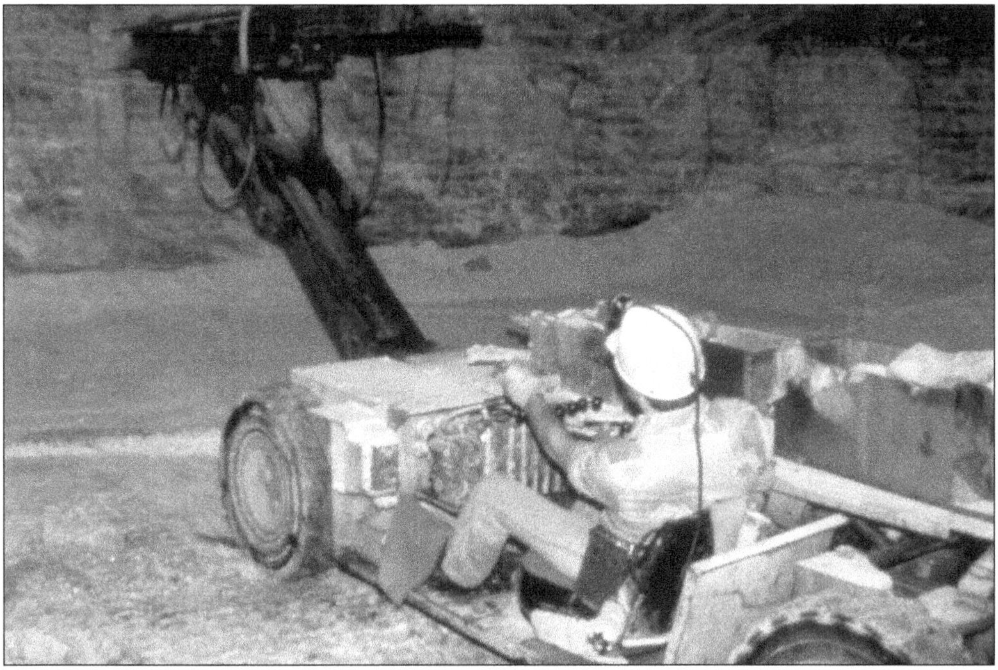

As with the undercutters, the newest drills are hydraulic like the one in this late-1970s image. They have one boom that can drill a room in only a few minutes. Since salt miners use the same equipment as coal miners, the drill is low to the ground to fit into tight coal seams even though the ceilings at the Carey Salt Mine range from 11 to 17 feet.

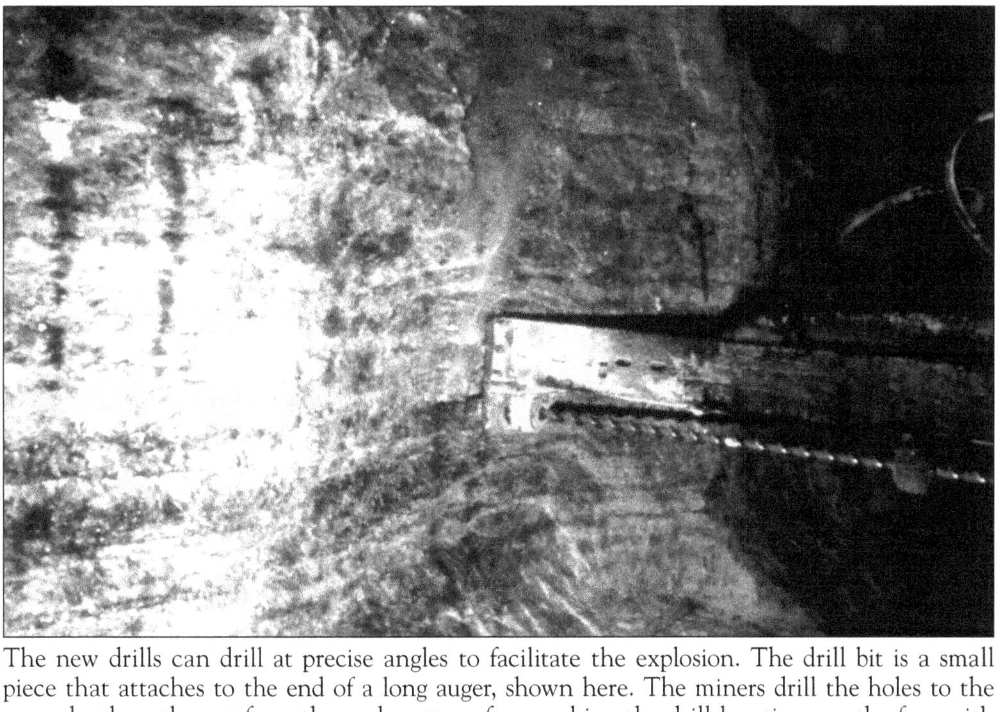

The new drills can drill at precise angles to facilitate the explosion. The drill bit is a small piece that attaches to the end of a long auger, shown here. The miners drill the holes to the same depth as the cut from the undercutter after marking the drill locations on the face with spray paint.

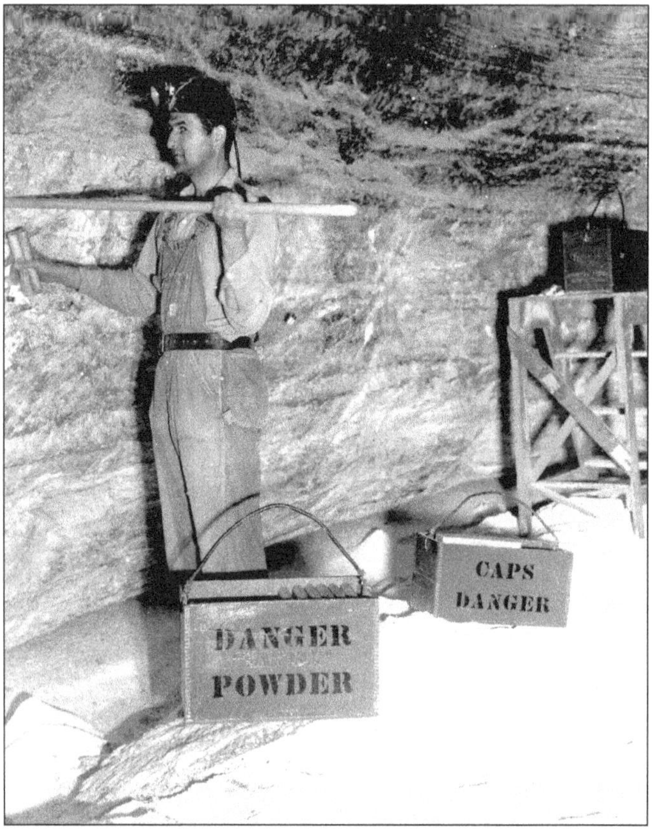

Once the undercutter and drill operators had finished, it was the powderman's turn. This 1951 photograph shows Vernon Horton tamping dynamite. Wire with electric blasting caps then connected the holes. In early years, the powderman blasted the room after it was wired and the loaders began their job before the smoke had cleared. Now miners only blast at the end of the last shift to allow the smoke to dissipate.

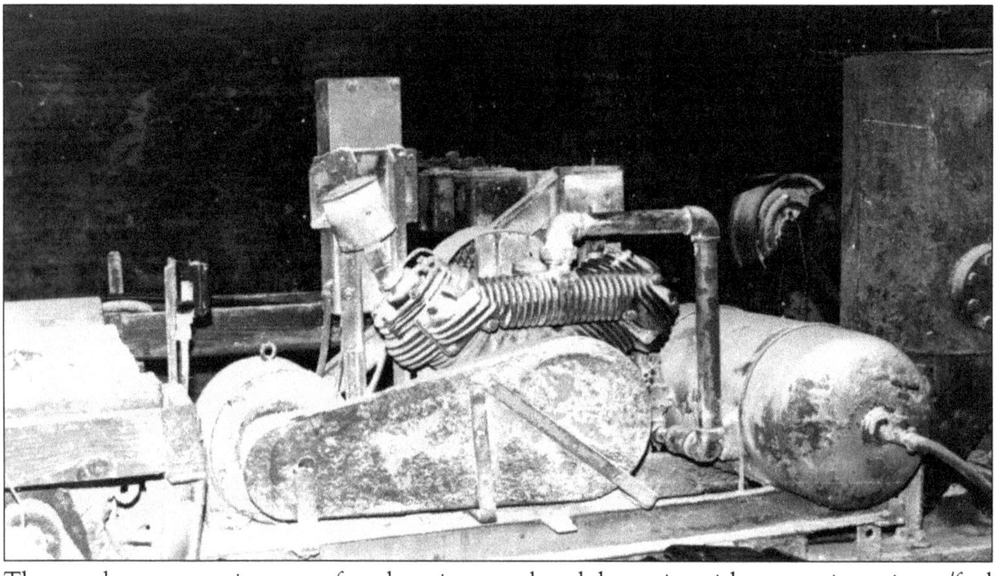

The powder car came into use after the miners replaced dynamite with ammonium nitrate/fuel oil (ANFO) in the late 1950s or early 1960s. The powder car had an air compressor (center) attached to the air tank (right). Out of range to the right sat a tank called a powder monkey that held the ANFO. The powderman used a pressurized hose to blow the ANFO into the holes drilled into the face.

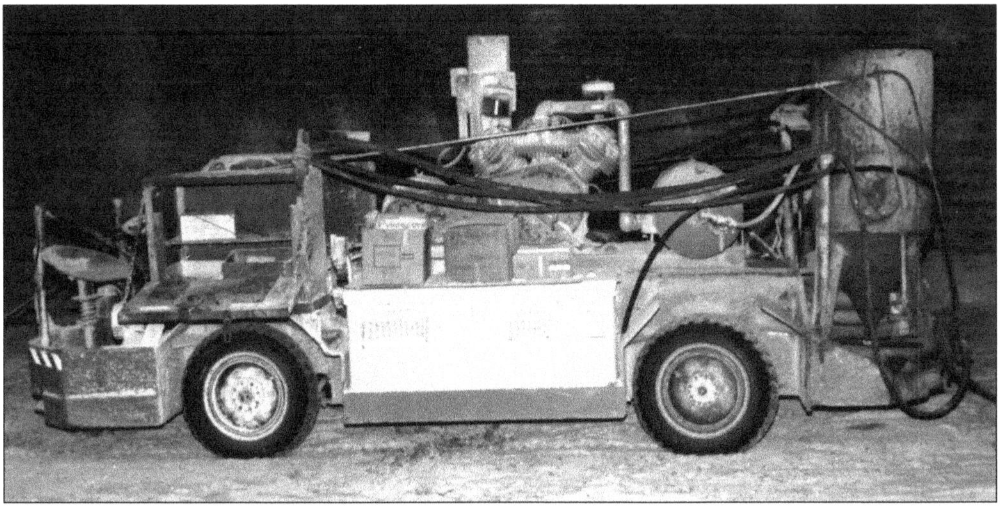

The Carey Salt Company purchased this Kersey tractor around 1977. The air compressor and tank are clearly seen in the center and right. The powder monkey is to the far right. The detonators were built with different time delays and had to be installed in a particular configuration across the face. Notes on the tractor's side indicate where to put the time delays when wiring both a 30-foot and 50-foot room.

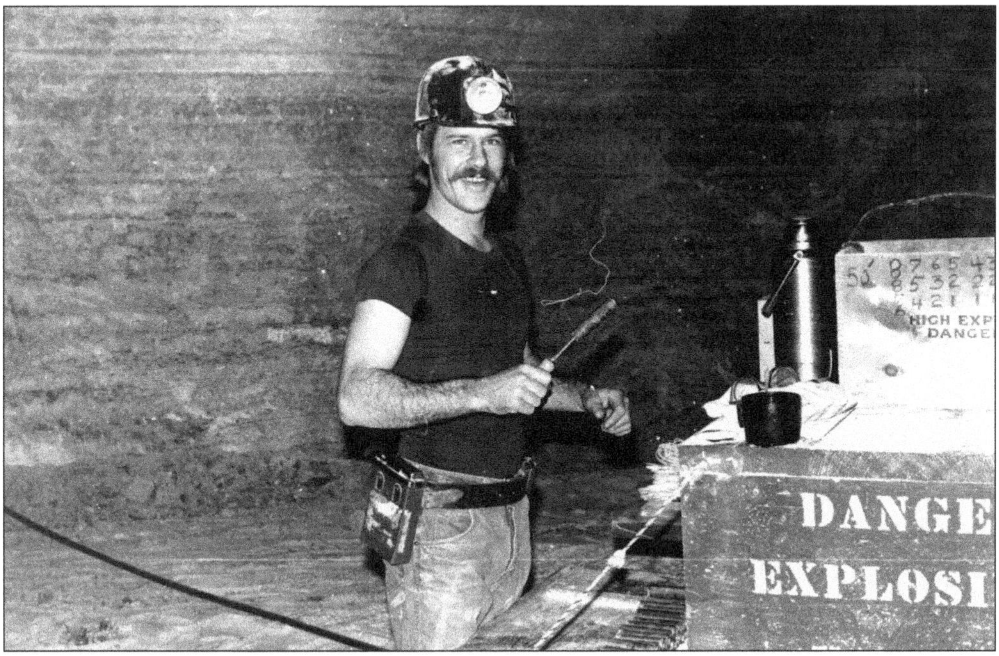

In this late-1970s photograph, powderman Dennis Ruckman pokes holes into dynamite sticks with a tool that is brass to prevent sparking. Each hole held a nonelectric delay detonator that was attached to the fuse that would be lit when blasting. The dynamite served as a booster to ignite the ANFO in each drill hole. In more recent years, the dynamite has been replaced by boosters that attach to the detonators.

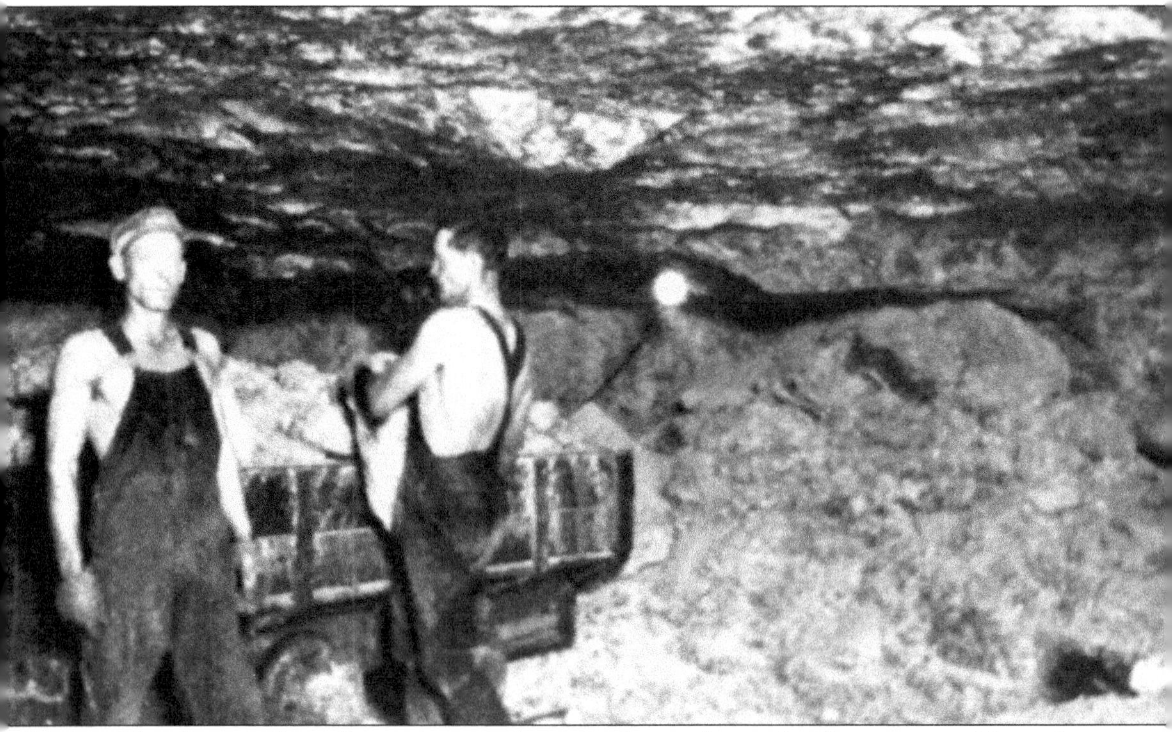

After shooting the wall, it was time to bring the loose salt topside. The majority of the earliest miners comprised the hand-loading crew, like the pair in this *c.* 1928 image. According to a memoir by miner Bennie J. Pallister, the loaders worked in pairs, using their own picks, shovels, and wedges to break the rocks before lifting them into the four-ton cars. Each pair would fill a car in about an hour and then present it to the locomotive engineer, who would give each of them a small brass chit to carry. The miner turned in his chits, which were each worth 50¢, daily to be recorded for his pay. A lost chit meant no pay, and the engineer had to replace the brass piece out of his own earnings.

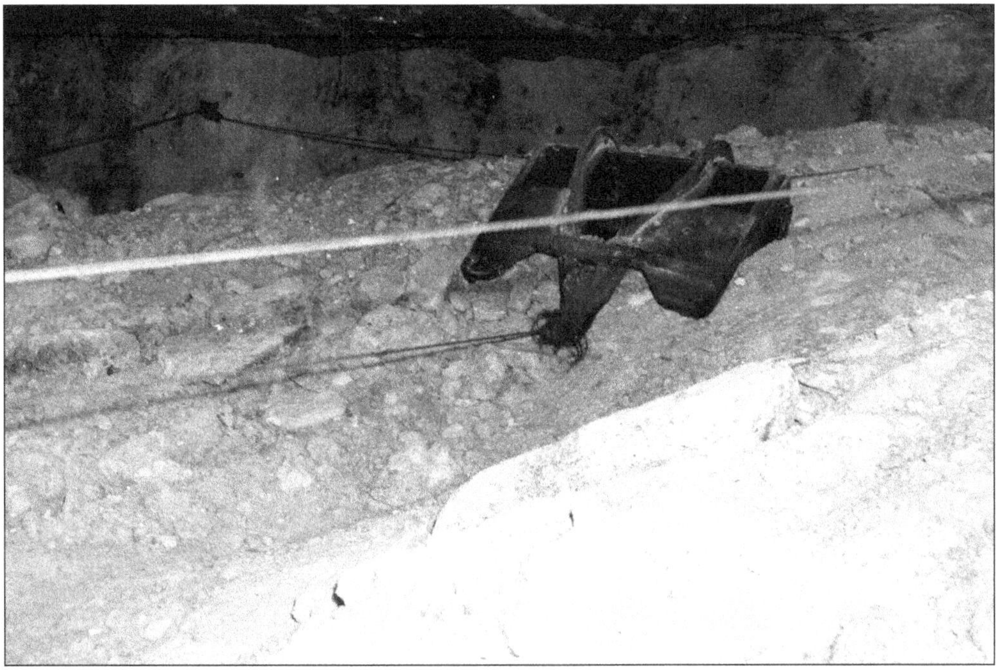

In 1923, mechanized loading was inefficient, though the technologies showed promise, and in the mid-1930s, the Carey Salt Company purchased scrapers to facilitate loading. According to Bennie J. Pallister, the scraper had "sheave wheels fastened in each corner of the face, and cables pulled through . . . to manipulate the scraper and push the salt up a chute to load the mine cars." This scoop in the Winnfield mine could have been used at Hutchinson.

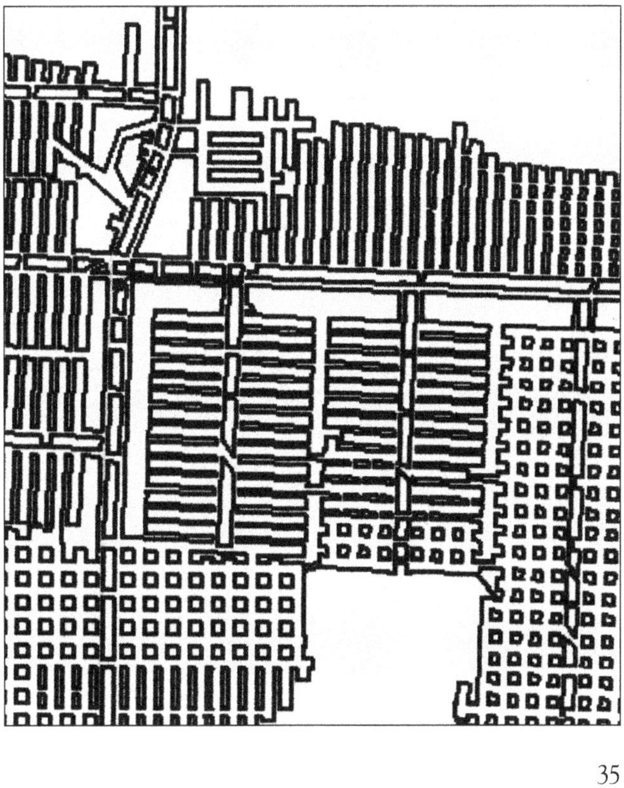

The scrapers and chutes caused the workers to alter the mining configuration. Abandoning the 300-foot bays, the miners created 50-foot spaces around square pillars. This yielded more salt without having to move the operations as often. This modern map of the mine shows the change in mining patterns at this time. The bays closest to the shaft are long, where the more recent pattern shows the 40-foot square pillars.

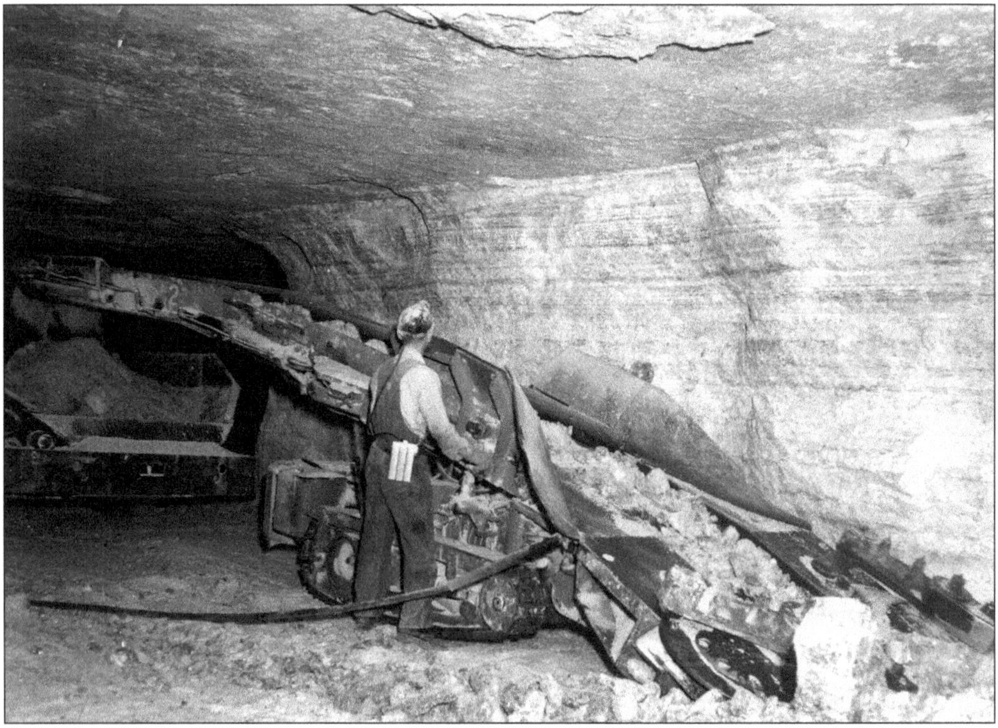

In 1941, a new system revolutionized the operations. An arm loader scooped the salt into a waiting shuttle car, which took it to a special loading station constructed at the rail line. This required fewer people to load the cars and less track to be laid since the salt could be loaded near the main line rather than having to lay track to the mine face.

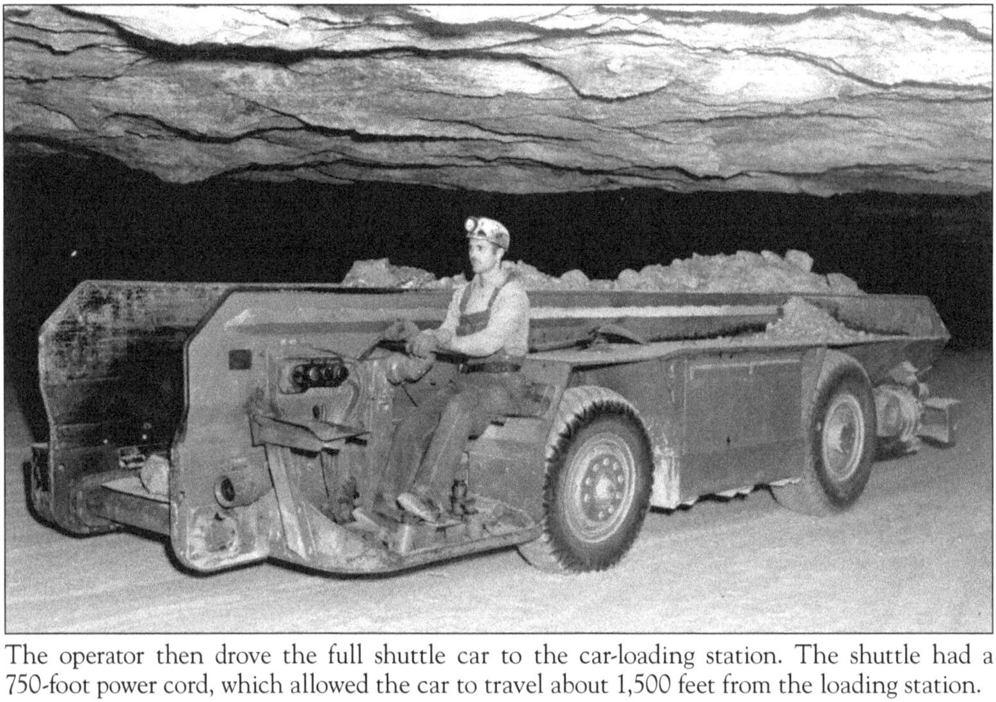

The operator then drove the full shuttle car to the car-loading station. The shuttle had a 750-foot power cord, which allowed the car to travel about 1,500 feet from the loading station.

The shuttle car impacted the way the mine was formed. The shaded areas on this map reflect the distance the shuttle car could go while tethered to its power source at the loading station. During the period of the arm loader and shuttle car, the mine caverns were blasted in a roughly circular pattern. These patterns disappear in the areas mined after 1983, when the shuttle cars were abandoned.

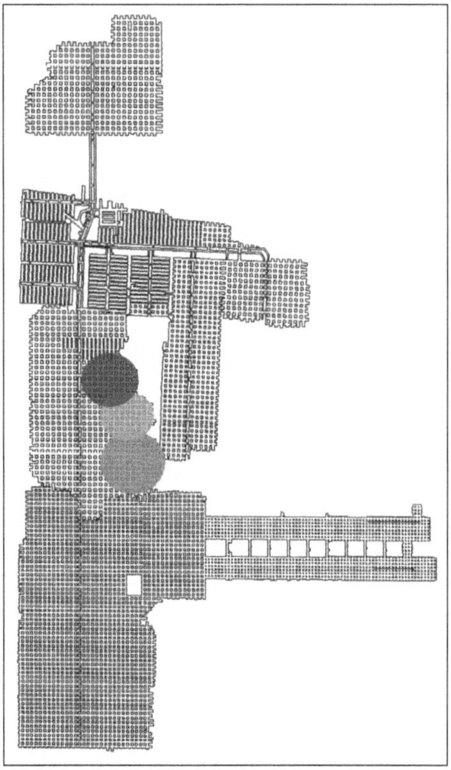

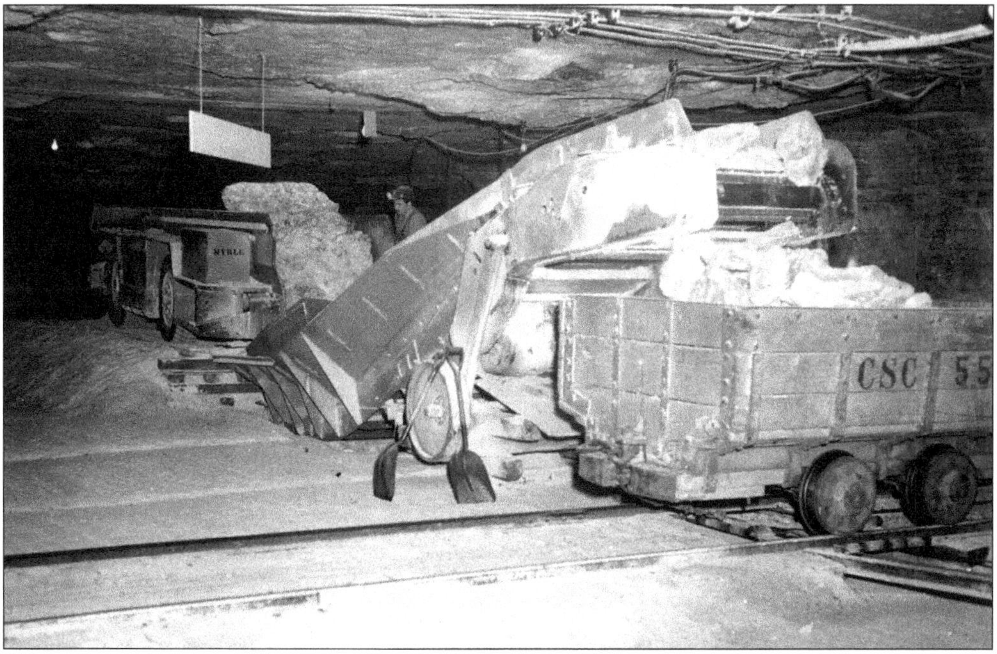

At first, the car-loading station was a simple setup. The shuttle car drove up a ramp that was braced with wooden cribbing and dumped the salt into a pit that held a flight conveyor. The salt traveled up the conveyor and dropped into the waiting ore cars on the track. Myrle Montgomery's first name can be seen on the shuttle car. He operated the shuttle car for several years.

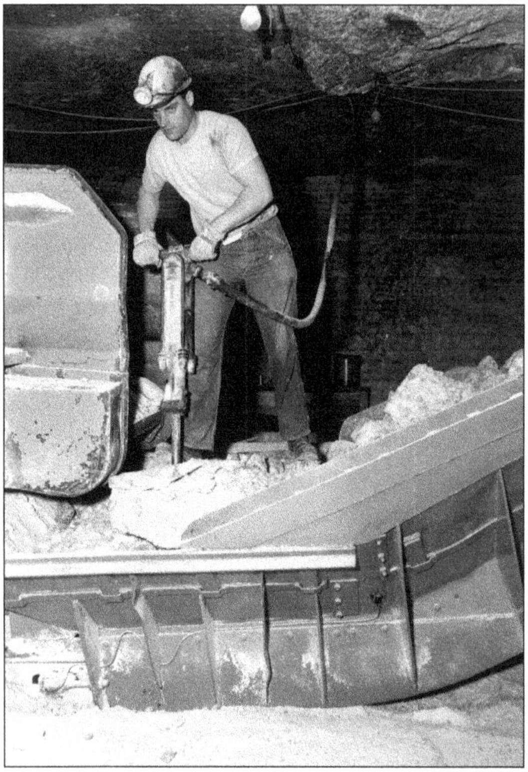

In the days of hand loading, the loaders broke the pieces into sizes they could lift into the car. These pieces were small since salt is heavy. Once mechanization took place, a miner broke the largest lumps into pieces as they came off the shuttle car. The largest lumps caused damage to the ore cars, rotary dump, and skips. This photograph shows Earl Bush at the jackhammer.

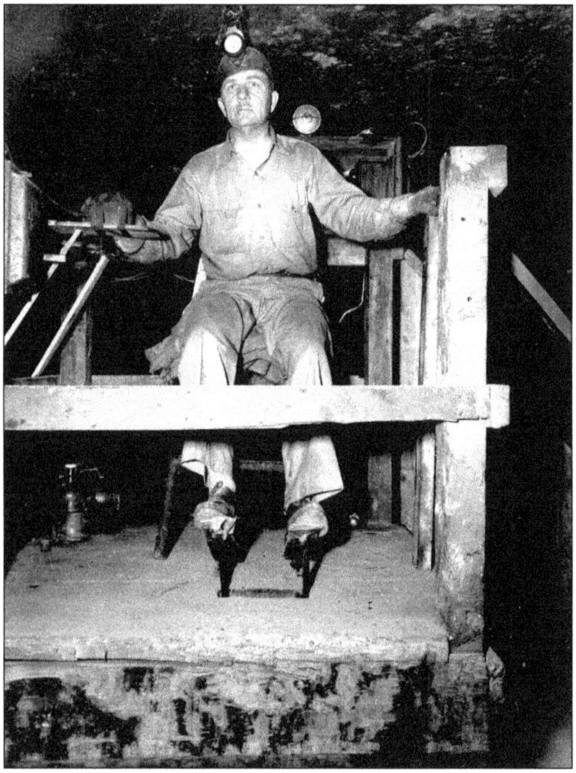

Roy Turnquist operates the controls to the car-loading station in this World War II–era photograph. The platform sat atop a pedal-controlled winch, which attached to cables connected to the first and last railcar. The winch moved the cars back and forth underneath the loader. Turnquist's hand operates the lever, which regulated the flow of salt into the ore car. His lunch pail sits in the background, with an enormous beverage jug.

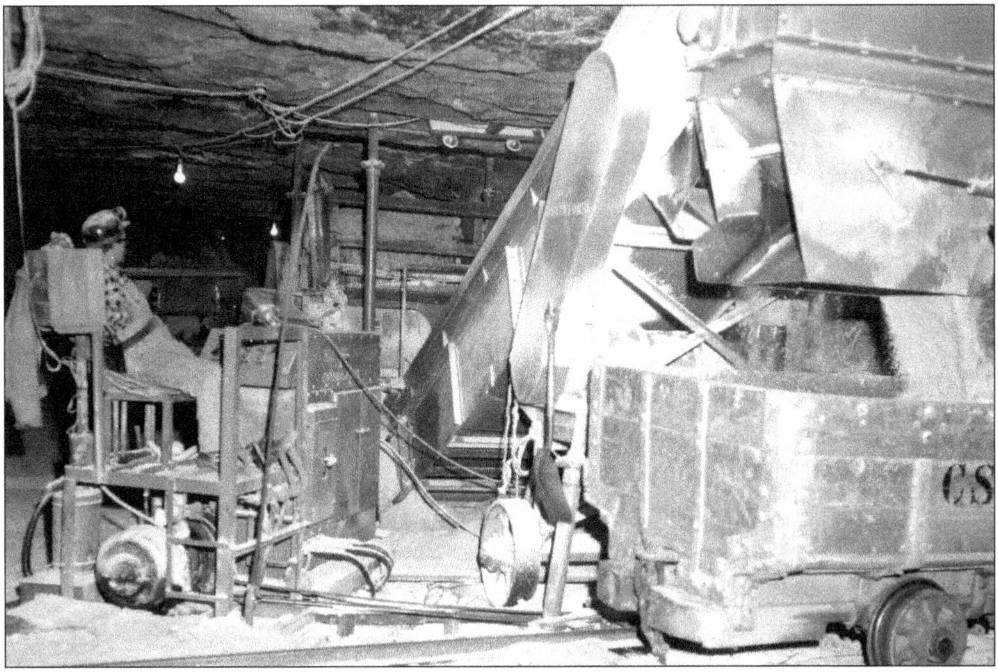

Around 1957, the miners added a crusher to the loading station to reduce the salt before loading it into the ore cars. The first flight conveyor dropped the salt onto the crusher conveyor, which then crushed the salt and loaded the cars. Mine officials were hopeful that the crusher would greatly increase production and reduce maintenance on the ore cars, rotary dump, and skips.

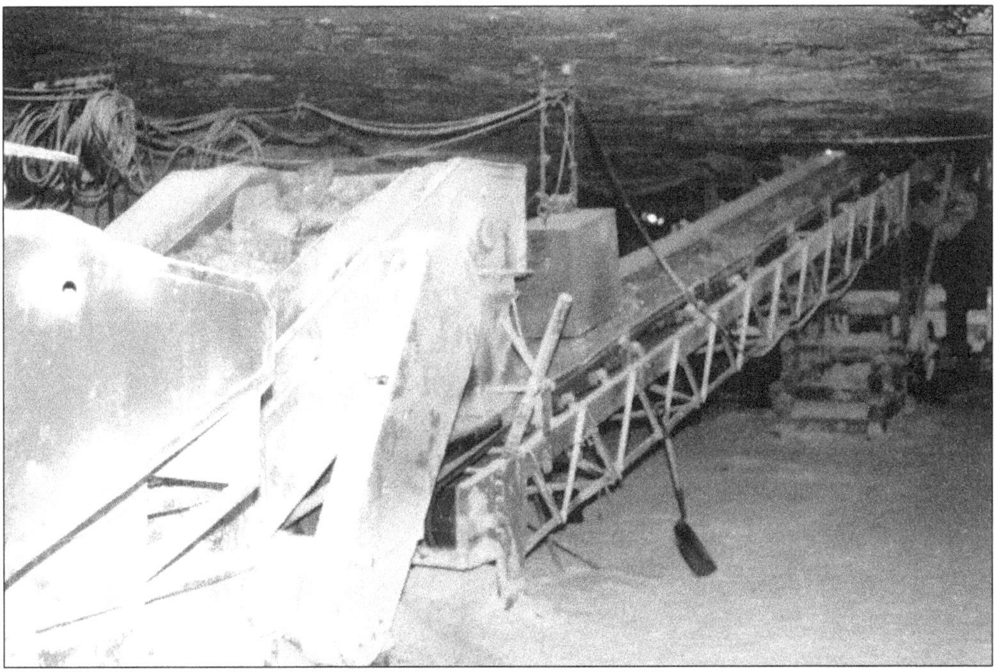

Later yet another section of the loading setup was added. Miners installed a large magnet over a final conveyor to remove pieces of wire and other debris swept into the shuttle car along with the salt. This photograph dates to around 1980.

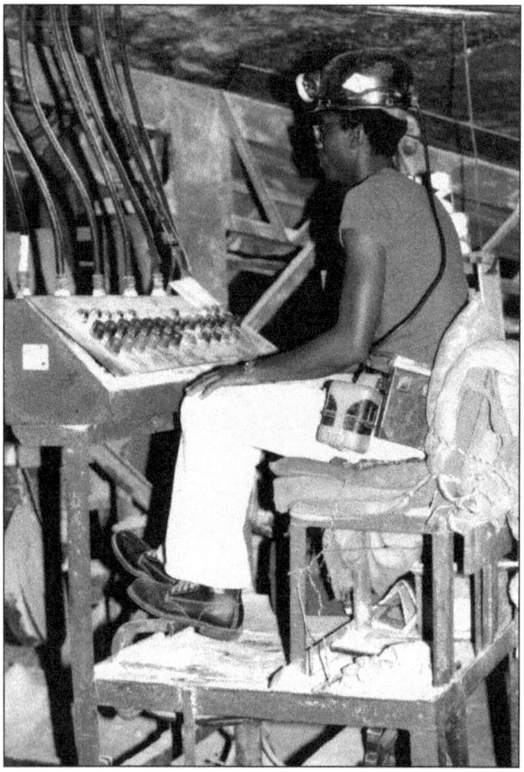

Mike Miller sits at the controls that ran the crusher and loaded the cars in this pre-1983 photograph. The buttons and pedals ran the different components of the crusher, from the drag chain that brought the salt from the shuttle dump pit to the gate that opened the chute to dump the salt into the cars.

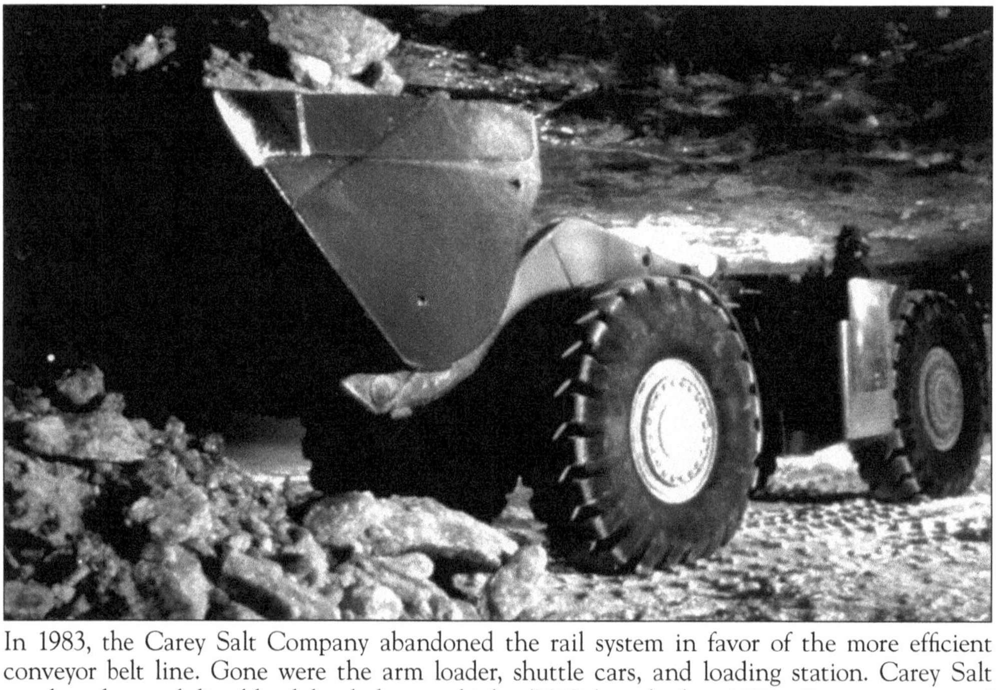

In 1983, the Carey Salt Company abandoned the rail system in favor of the more efficient conveyor belt line. Gone were the arm loader, shuttle cars, and loading station. Carey Salt purchased several diesel load, haul, dump vehicles (LHDs) in the late 1980s. Operators scooped six tons of salt from the face and drove it to the tail of the belt line, where it would begin its automated journey to the hoist.

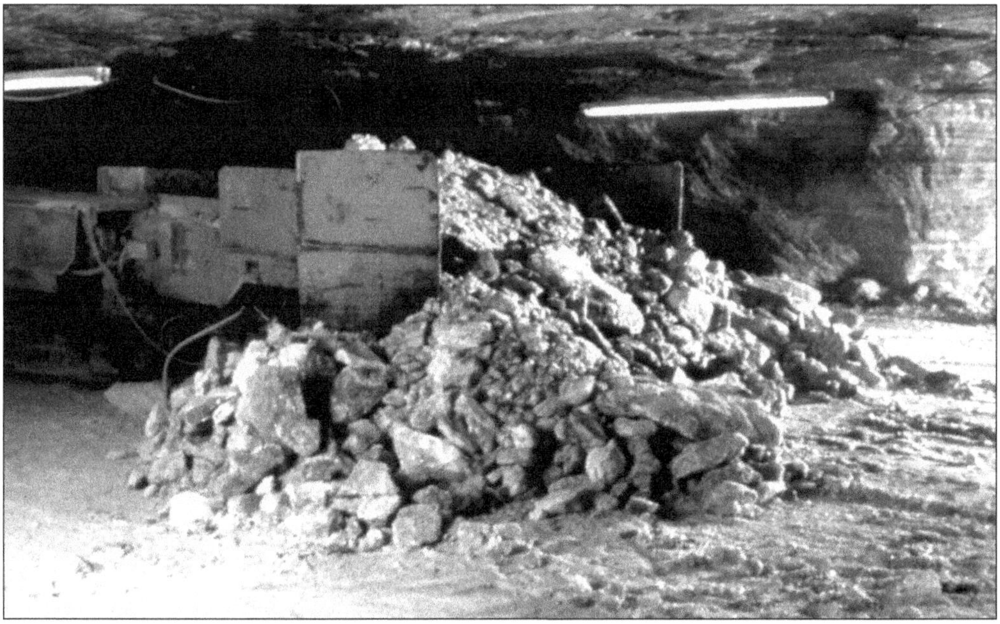

The initial crushing process remained underground. Atop the end of the belt sat this crusher, which reduced the salt to softball-size chunks. In this photograph, the LHD has dumped the salt into the crusher to the right. To the left begins the belt line where the salt will travel to the hoist, eventually passing through a second crusher to reduce it to gravel. The Hutchinson Salt Company still uses this equipment.

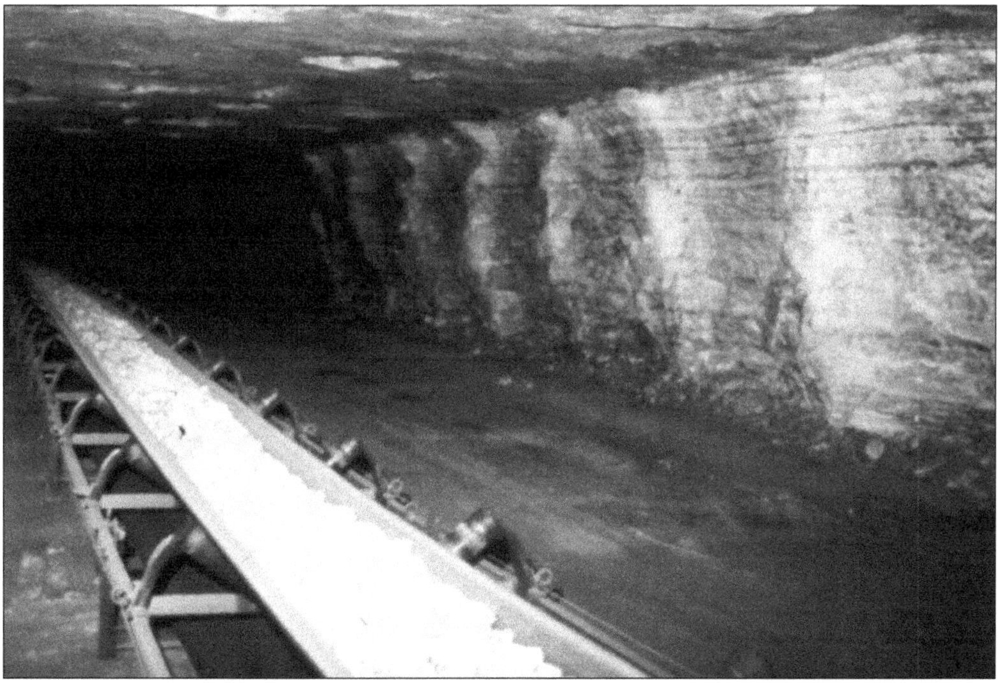

Then as now, the conveyor belt consisted of several sections of line ranging from 1,200 to 2,400 feet long. Where one belt ends, the salt drops onto another belt to continue the journey. Several lines are used at this mine because the belt must turn so many corners.

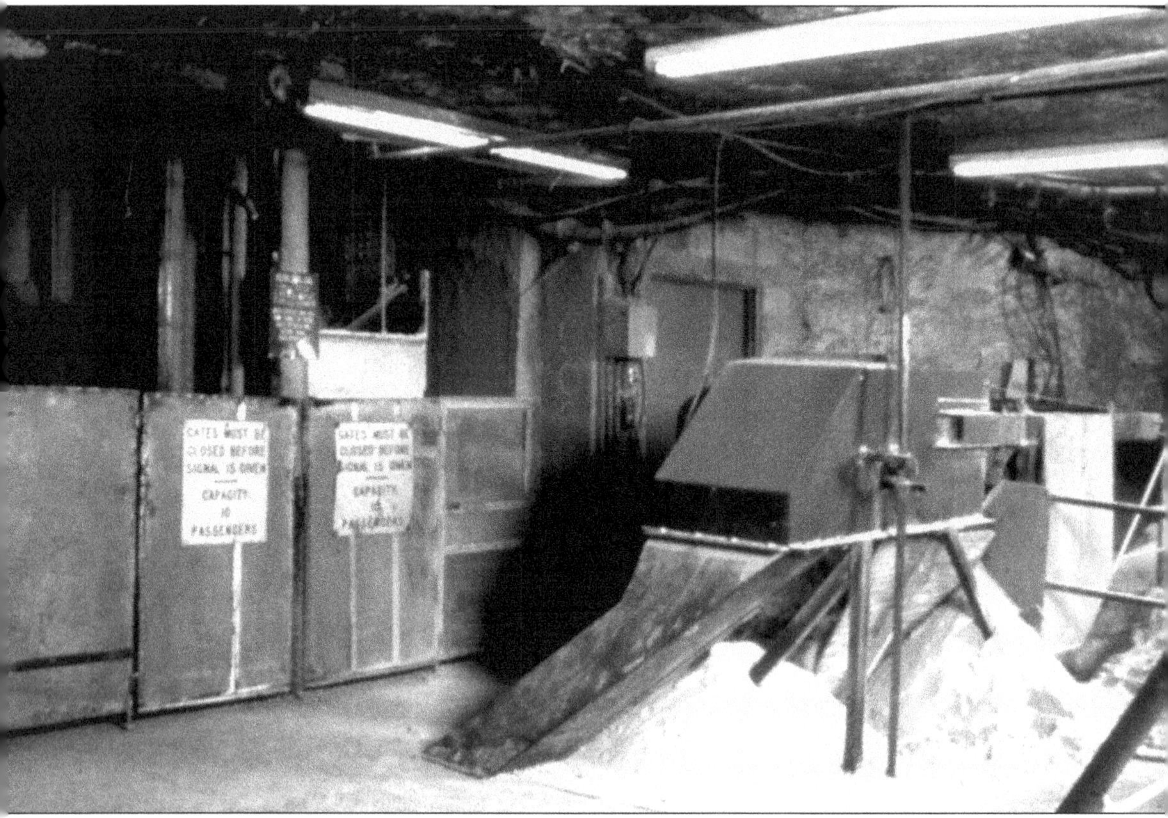

The belt line eliminated the rotary dump as well. The hoist operator merely channels the salt through one of two chutes that lead to either skip. This photograph shows the end of the belt line next to the hoist.

Four

The Mine Environment and Maintaining the Mine

The mine is surrounded by salt. The environment is so surrounded by salt—ceilings, walls, floors, and dust so thick it can look like snow—it is easy to forget that the mine is actually a man-made thing. Everything underground—the spaces the miners carve, the air they breathe, the light by which they operate, the equipment they use—has been imported from the surface. The mine is not a natural thing, although it is embedded in nature.

Mineral rights agreements with landowners aboveground dictate the direction of the miners' progress underground. Engineers carefully plan the direction that the miners will blast before any mining takes place. Workers survey and map the caverns to make sure they are straight and at the correct level.

The air, which is forced into the back of the shaft and exhausted out the front of the shaft, is protected by the miners themselves and by the Mine Safety Health Administration (MSHA). Diesel particulates from the equipment are monitored and regulated. Air is channeled through a path to keep it flowing to the face and back.

If air keeps the mining crews working, electricity keeps everything else working. The maintenance department wires and rewires lines to the ever-moving face to keep light where it needs to be and the equipment plugged in. It is also responsible for maintaining and repairing the vehicles and equipment.

As mines go, salt mines are considered to have one of the best environments for miners. The Carey Salt Mine's constant 68 degrees and low relative humidity at the face makes the work bearable, and the salt seam—at least since the 1940s—ranges between 11 and 17 feet high. Miners can stand up straight. Salt mines have no methane or other gases that lurk underground, ready to explode or asphyxiate. This is not to say that salt miners have no worries. Underground caverns are prone to roof falls, and the task of mining itself has cost some men their lives.

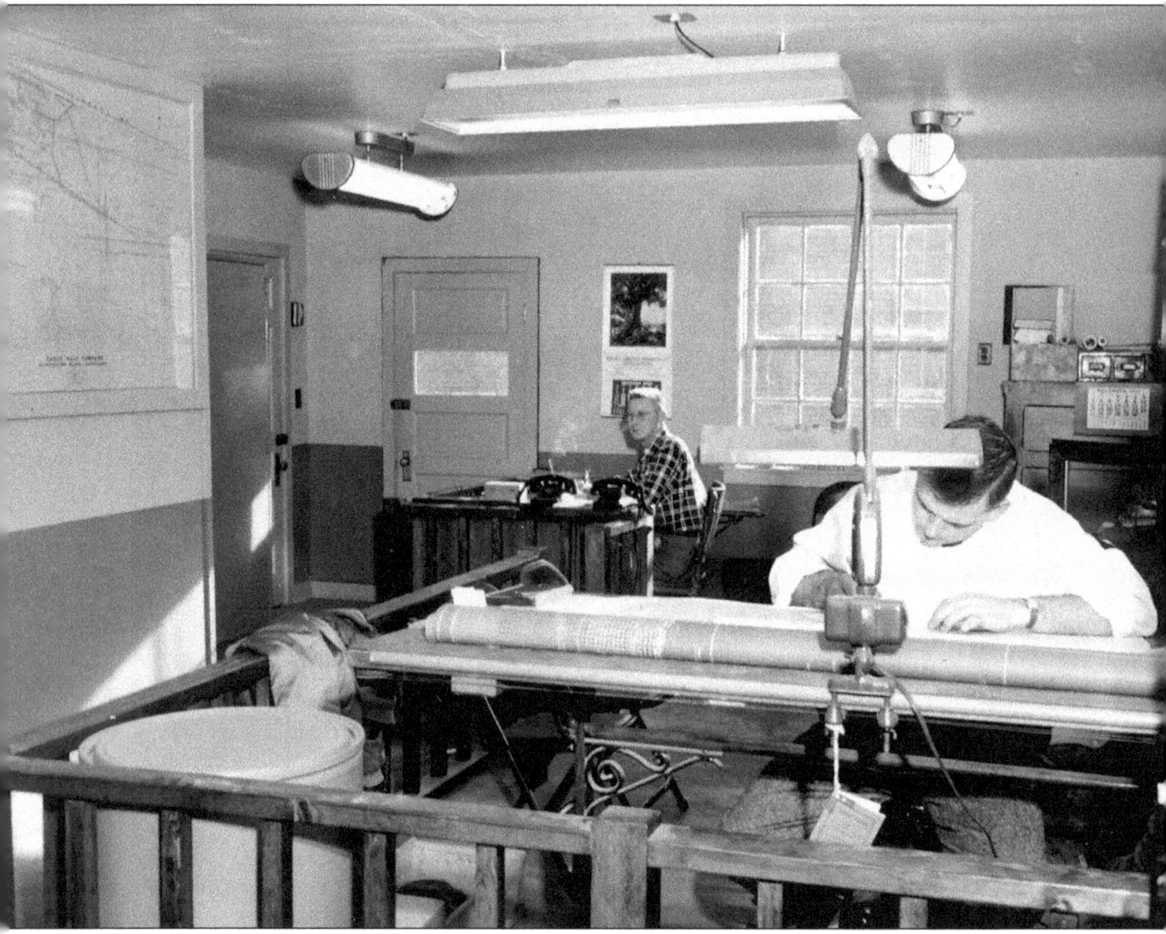

The Carey Salt Company employed engineers who worked on both the evaporating plant and the mine. They designed and improved equipment, made calculations regarding the speed and efficiency of the process, and mapped and determined the configuration of the mine.

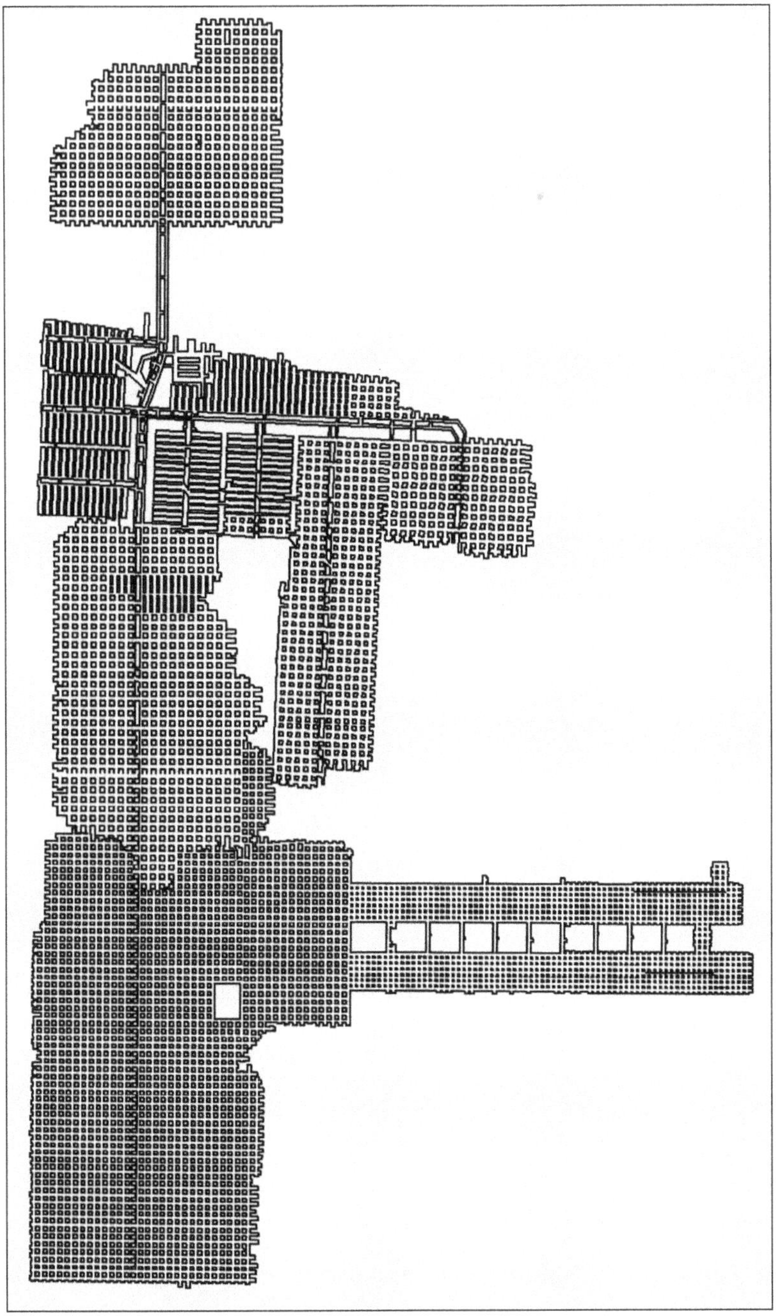

This 2007 map of the mine reflects the different mining technologies as well as outside forces at work on the mine. Salt quality can impact the mine layout, as shown by the large block of unmined salt in the center of the map. Mineral rights also play a large part. In 1956, the Atchison, Topeka and Santa Fe Railroad (ATSF) denied Carey Salt the right to mine underneath its property. It did, however, allow it to blast a tunnel to the north to access areas where it did own the mineral rights. Accurate mapping is critical to the mining process because other entities, such as gas and oil wells, are also invading the space underground, as shown by the large solitary pillar of salt toward the south of the mine.

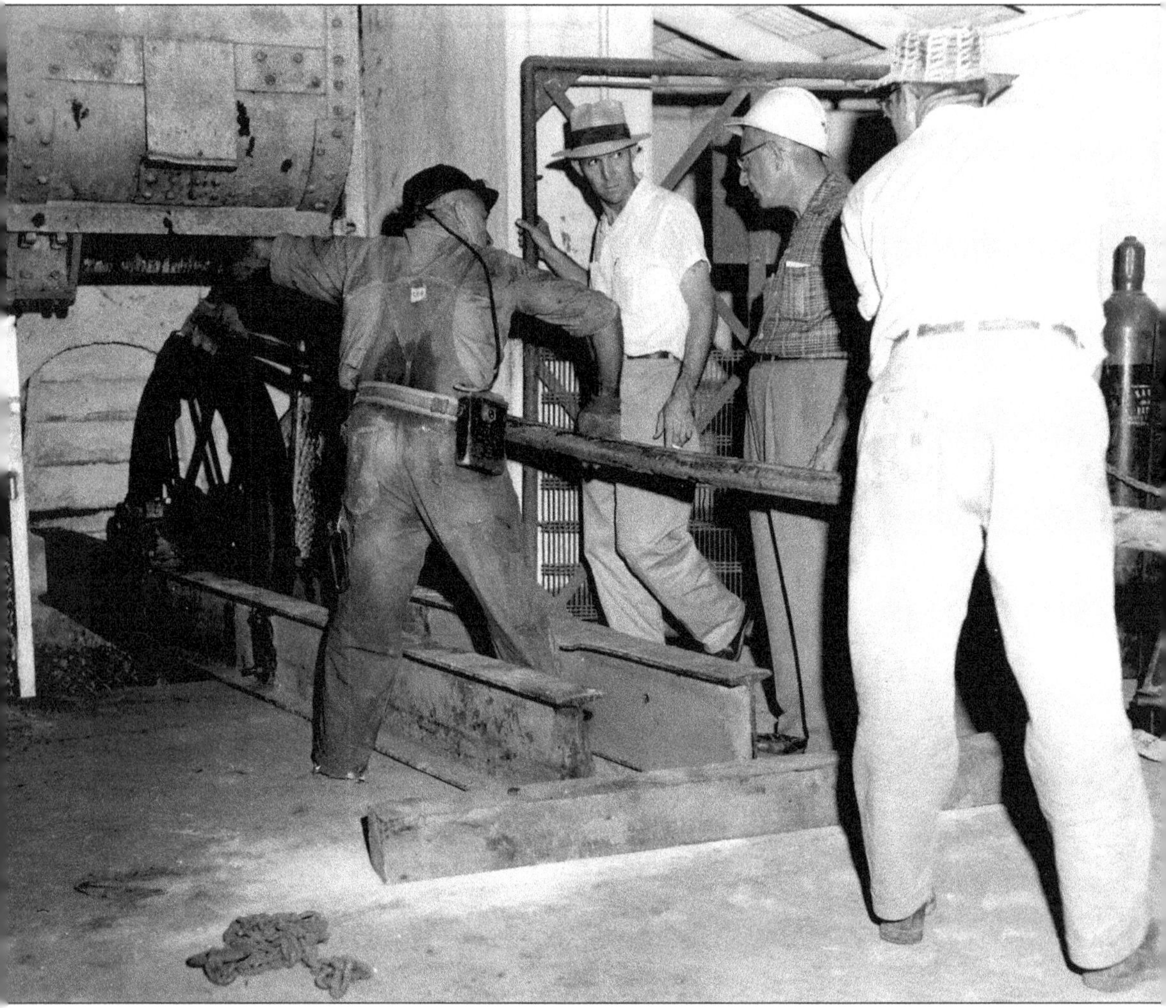

The July 1957 report to the board of directors states, "At the present time we are in the process of removing the small electric cable installed in 1923 and the old telephone cable, which failed some years ago, from the shaft proper. A new and larger cable of more capacity is being installed in the shaft. The new cable is of such size to handle any underground expansion and mechanization that might be necessary for a number of years to come." The new cable, which carried alternating current, was painstakingly lowered down the shaft with a pulley resting on two I beams. In the next directors' report, Steven B. Horrell remarks, "From the condition of the old cable it appears that this work was done none too soon."

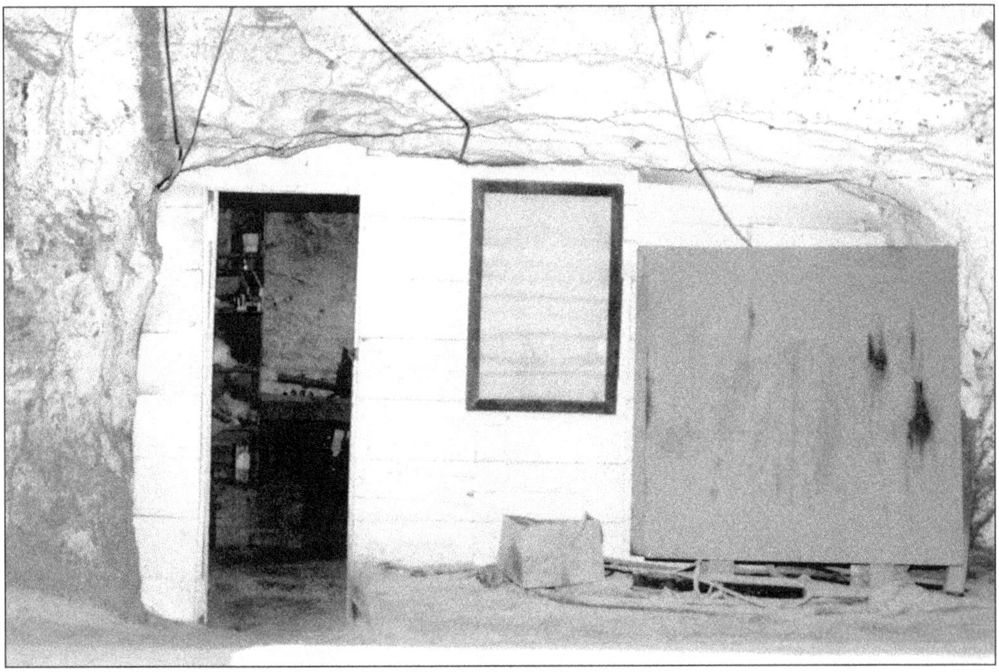

The foreman's office sat near the hoist underground to make him available to supervise the operations. One of his other jobs was to sell tools to the employees, such as picks, shovels, safety shoes, gloves, and other accessories. For a long time, the foreman's office was the only place in the mine that had a telephone.

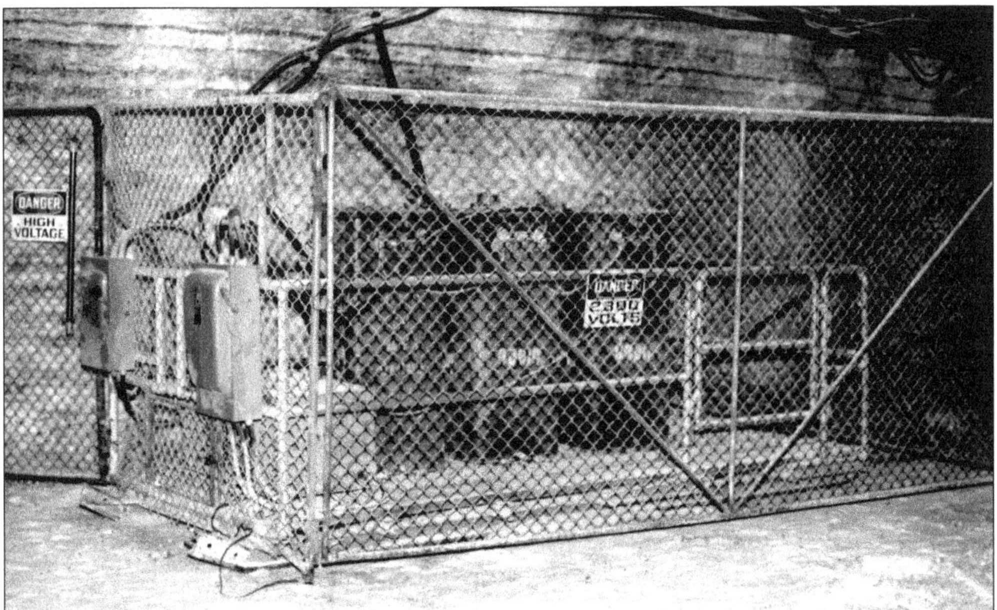

Underground transformer stations changed the high voltage of the main power line to a smaller 480 voltage. Mining equipment like the drill and undercutters all operated on a three-phase alternating current system. The locomotive trolley line and the shuttle cars were converted to direct current.

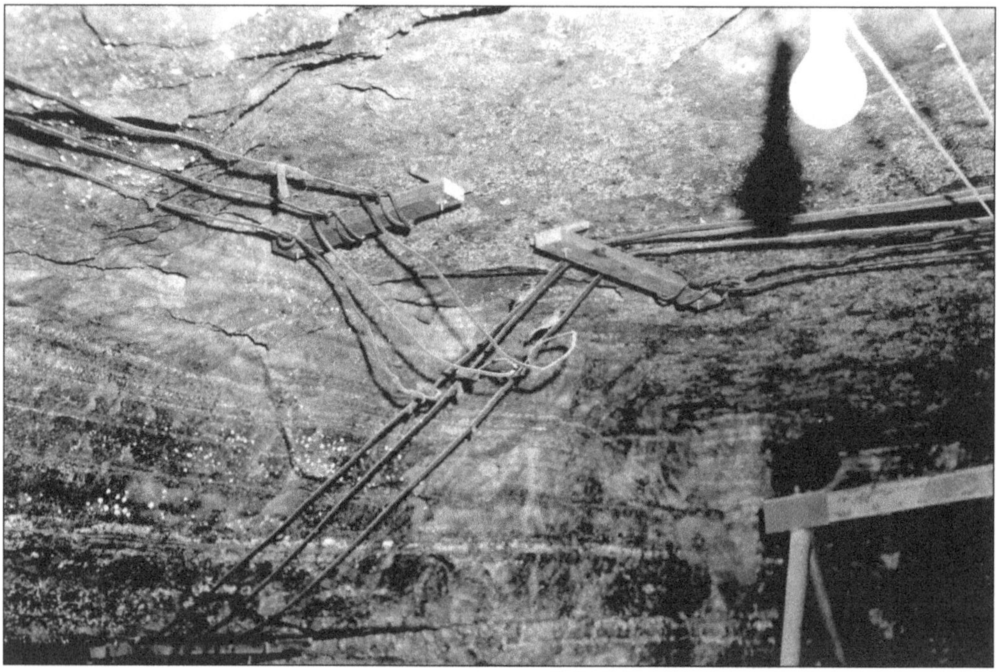

Three 480-volt power lines were strung parallel to one another from the transformer to the mine face. This 1957 photograph shows the connection of two strings of power lines. A stray bird had somehow entered the mine and can be seen perched on the connection on the right.

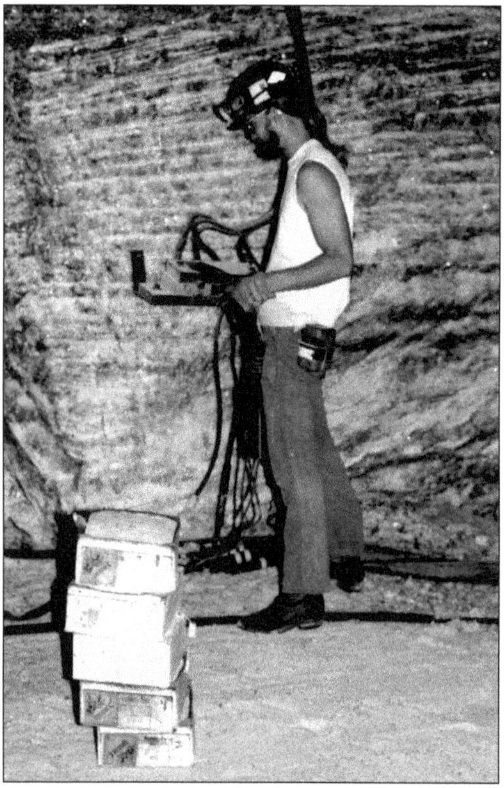

Each power line was required for the three-phase system, which was more cost-effective than a single-phase system. This late-1970s photograph shows powderman Bob Haskins plugging his equipment into a three-phase receptacle mounted to the wall. If the leads were not placed in the correct order, the machine would run backwards. Change any two cords and the problem would be corrected.

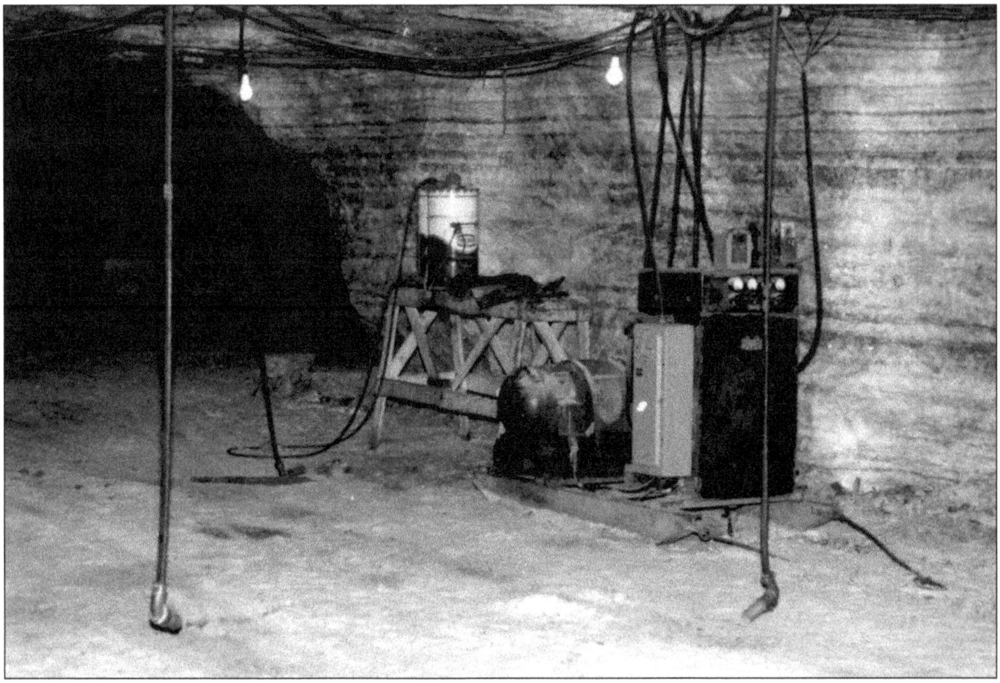

Two of the shuttle cars in the 1970s and 1980s were battery operated and had to be plugged into this charging station at the shift's end. The motor-generator (M-G) set along the wall created the energy necessary to recharge the batteries.

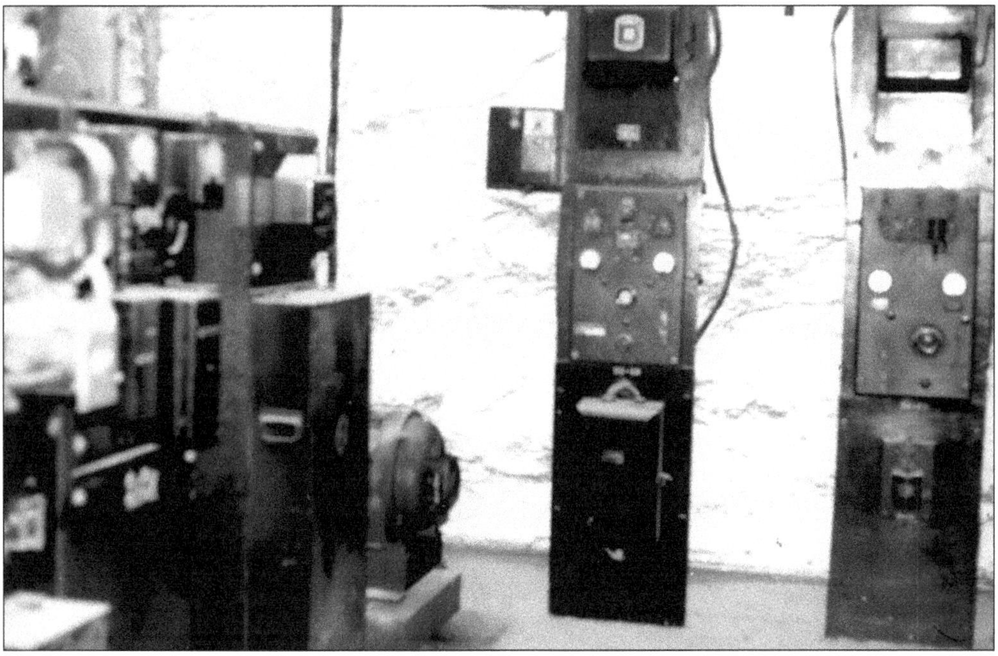

The large locomotive was also charged at the end of the shift, at a station near the hoist. Originally it was plugged into the M-G set. The squat, round motor sits in the corner in this photograph, and the tall, oblong structure (center) is the generator. Later the train plugged into a rectifier box and the M-G system was abandoned.

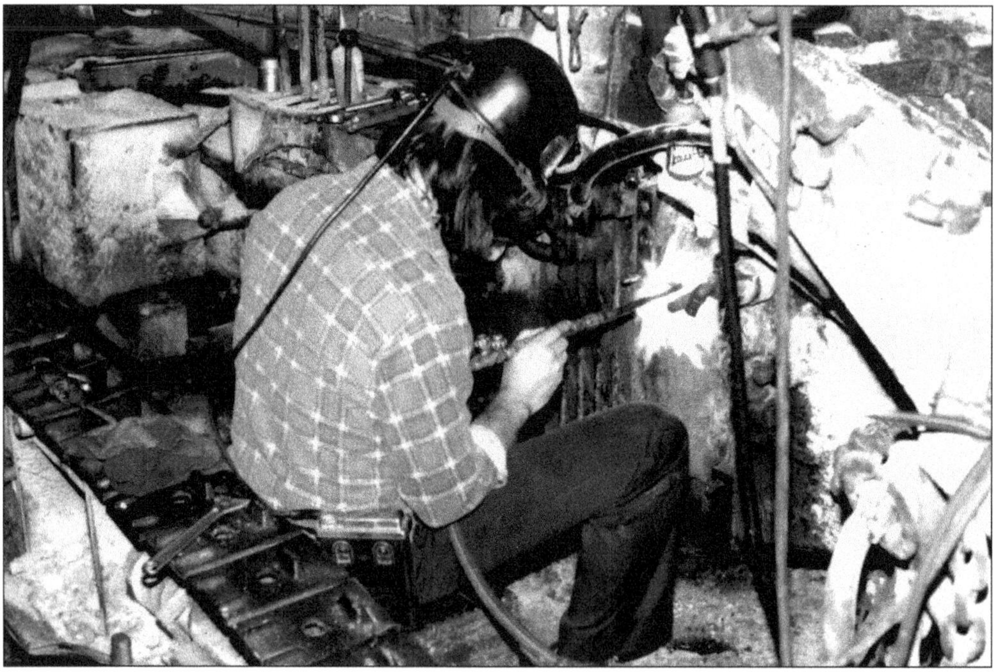

The maintenance workers kept everything running smoothly. In this late-1970s photograph, Marty Gumble repairs the arm loader at the mine face.

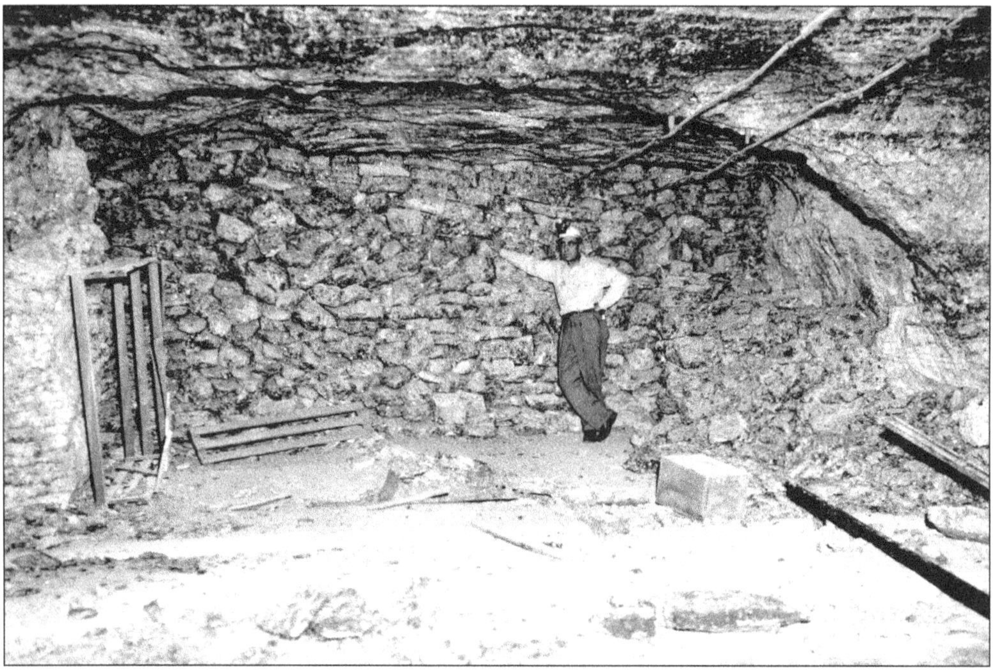

Airflow is critical in a mine. To channel the air from the shaft to the face and back again, miners built walls like this one in the Lyons mine. The rock walls were not airtight. In at least one Hutchinson location, the miners covered the rock with a lime-stucco mixture to seal it. Other walls were made of used dynamite boxes and, later, wool and vinyl curtains called brattice cloth.

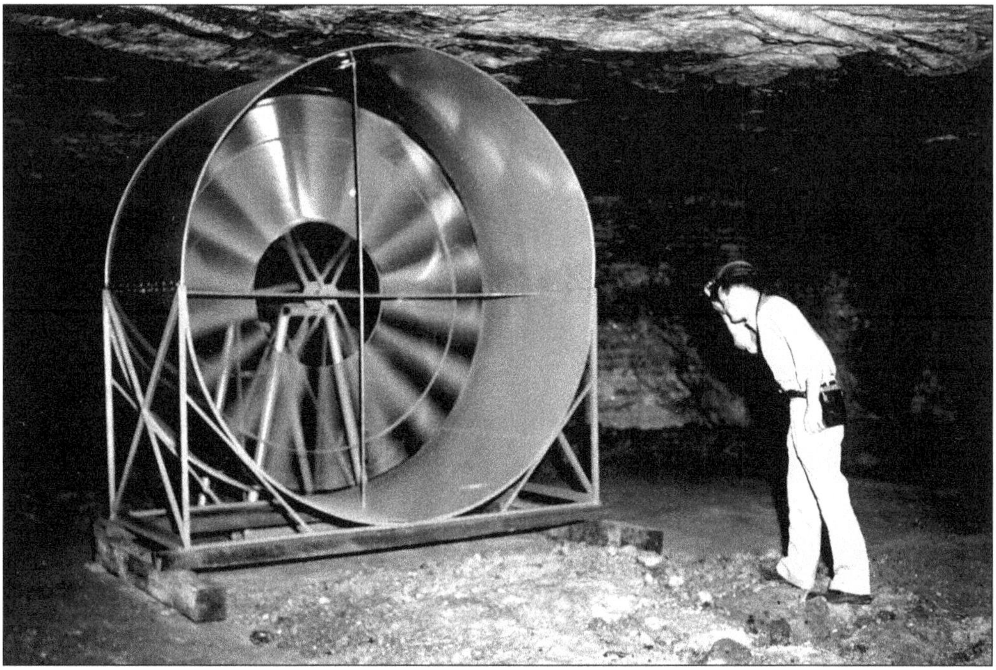

An unidentified mine supervisor inspects the fan that had been placed in the mine to improve the flow of air in this 1955 image.

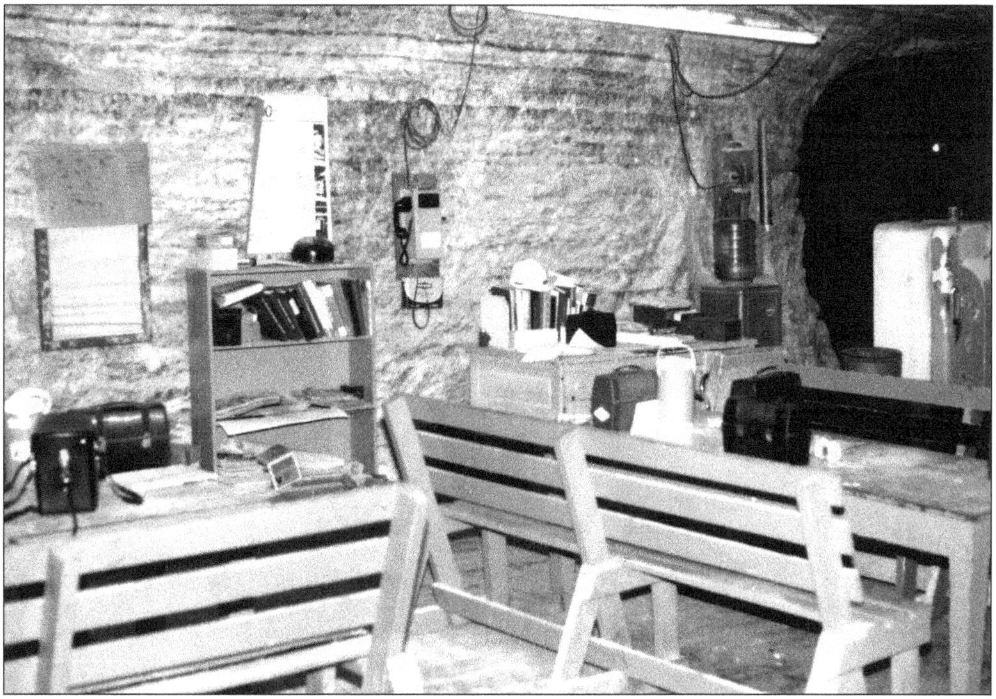

For many years, the miners ate lunch at informal break areas near the loading station, but in more recent decades, the miners had a designated break room. MSHA requires that the break areas come first in the line of exhaust. This would make them upwind from any likely fires that could occur, and they would still receive fresh air.

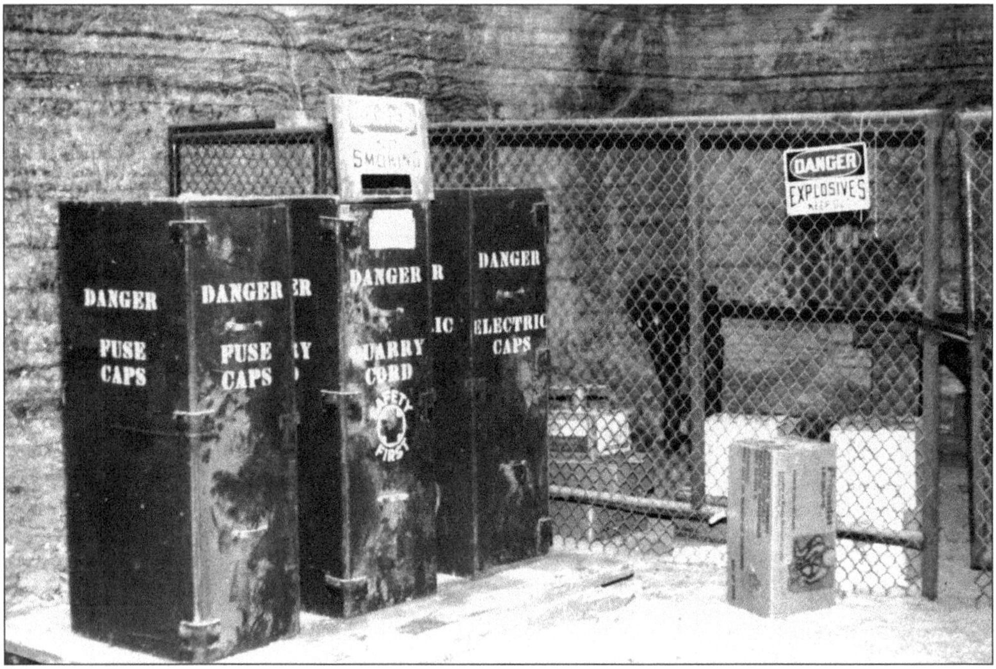

The magazine, shown here in this late-1970s photograph, holds the blasting supplies in a secure area. In recent decades, MSHA has required that the magazine be located last in the line of exhaust to ensure that any explosion would ventilate directly out the mine. This way, the miners, who would be positioned upwind, would still receive fresh air.

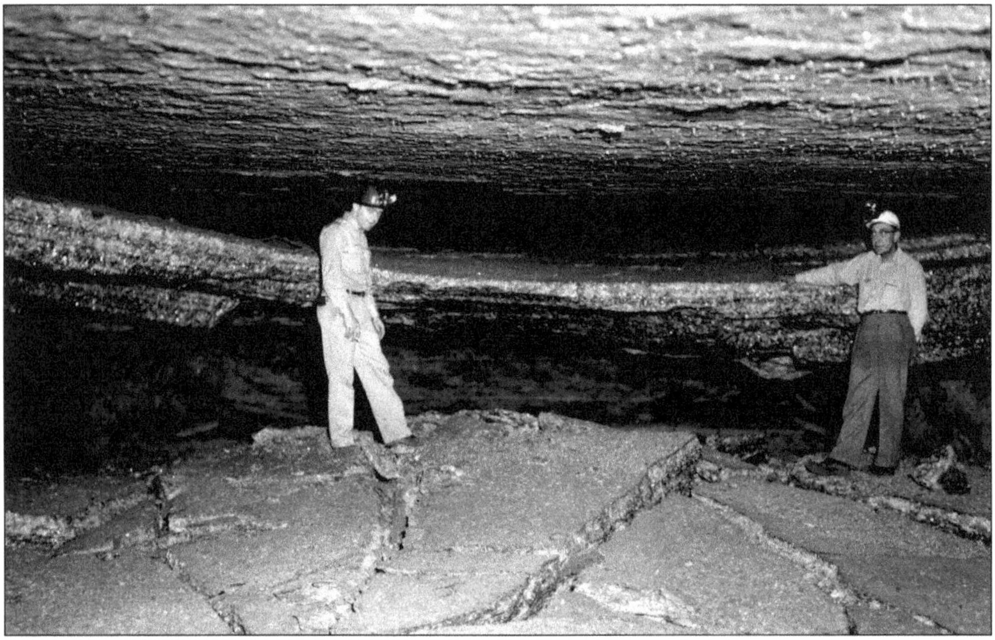

These unidentified men are standing on what was once the ceiling in the Lyons mine. Salt separates from the shale above it when the layers are weak and the pillars cannot support the overburden—the rock overhead. To eliminate danger, miners shoot or knock roof sags like this one to the floor before the salt can fall on its own.

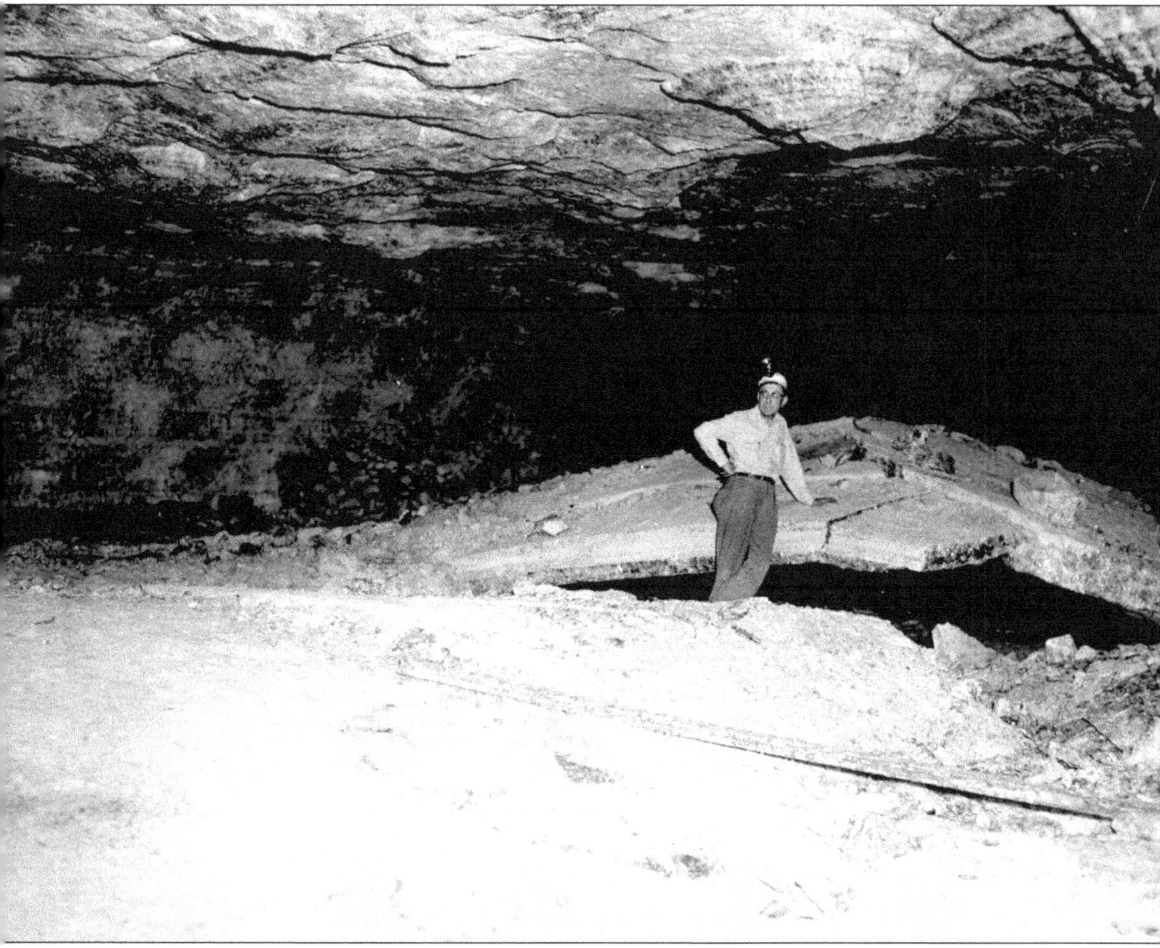

Sometimes the ceiling is stable, but the layers of salt underneath are not strong enough to support the pillars pressing on them. When this happens, a floor heave like this one from the Lyons mine occurs. Generally a floor heave and a roof fall will not both occur in the same place.

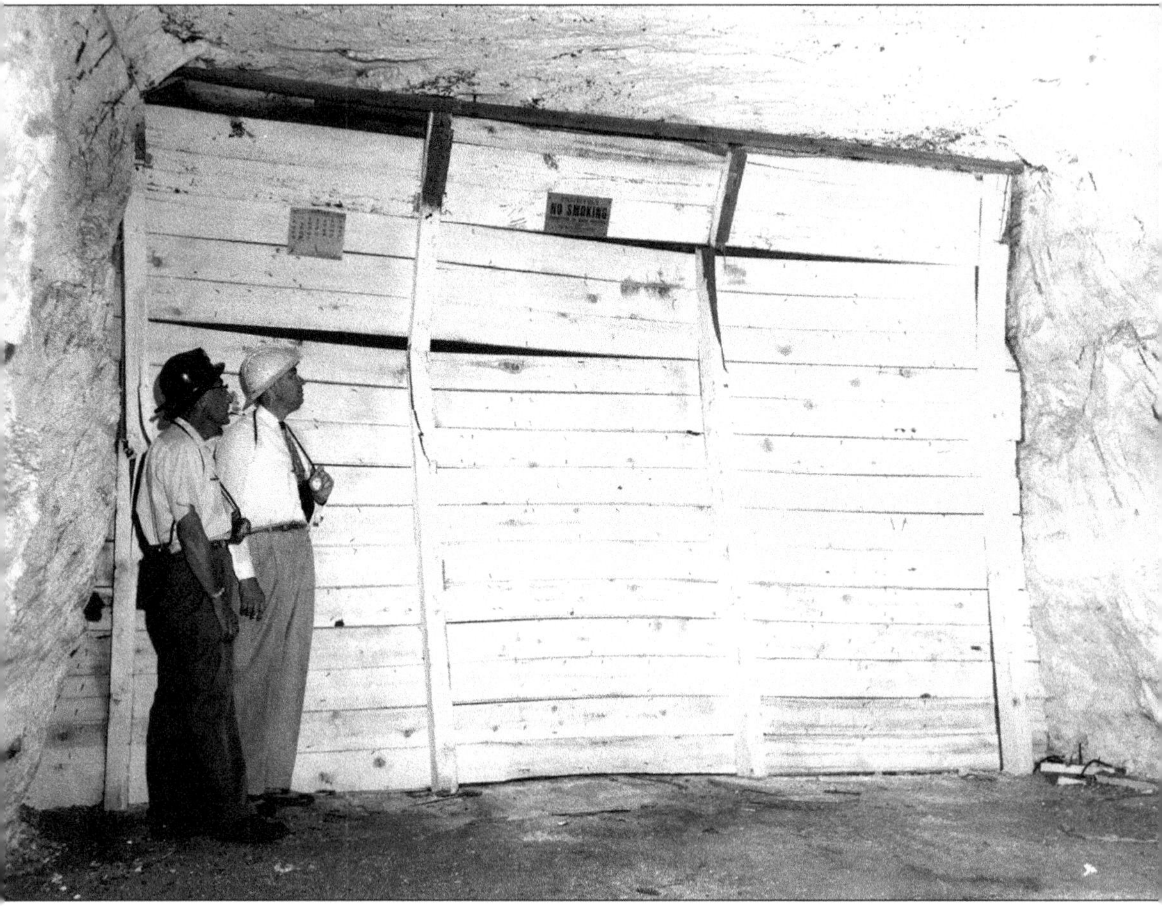

The wooden gob wall is buckling under the sagging overburden in this 1956 photograph. This image was shot for the Richardson Bass Company, an engineering firm investigating the possibility of storing liquid petroleum in old brine caverns. This firm conducted tests in the Hutchinson mine to establish that the permeability of mined caverns were different than in the brine wells. Mine manager Leo Reid (left) and Carey Salt Company chemist Paul Imes, shown here, appeared before a United States court hearing in Washington, D.C., to support the findings.

Five

Processing the Salt

Mining the salt is only the half of it. Blasted salt ranges from large boulders to powdery salt fines. To make salt desirable, the largest lumps had to be crushed and the sizes had to be separated into consistent sizes. Once the salt had been retrieved topside, the workers at the mill crushed and separated the different sizes into grades, storing them in bins until they were sold. Originally four grades were established of varying coarseness.

All the crushing occurred in the mill until 1957, when the portable crushing station was installed by the loading station underground. This simplified the process aboveground. In the mill, the salt was first crushed before it went through a series of screens to separate the grades, being crushed again if it was still too coarse. The process aboveground, like belowground, was a simple one, but the scale of the operations could make the tasks challenging. The vice presidents' operations reports throughout the 1950s complained that crushing created a bottleneck in the process because of the large volume of salt going through that needed to be recrushed for one reason or another.

Once the grades were established, the salt was packaged—usually in sacks of varying sizes—or sold in bulk directly to industries. The same salt was sold to companies in many different industries. In the first decades, the Carey Salt Company offered a wide variety of products to farmers from both the evaporation plant and the mine. The mine produced loose rock salt for mixing into their livestock feed or sprinkling between layers of hay to preserve it. Many mixed orders of pressed salt blocks from the evaporation plant and rock salt left on trains to rural areas all over the Midwest. After World War II ended, the growing highway system and increased use of automobiles put salt in great demand for ice control and road stabilization. These two industries have been a mainstay for the mine for decades.

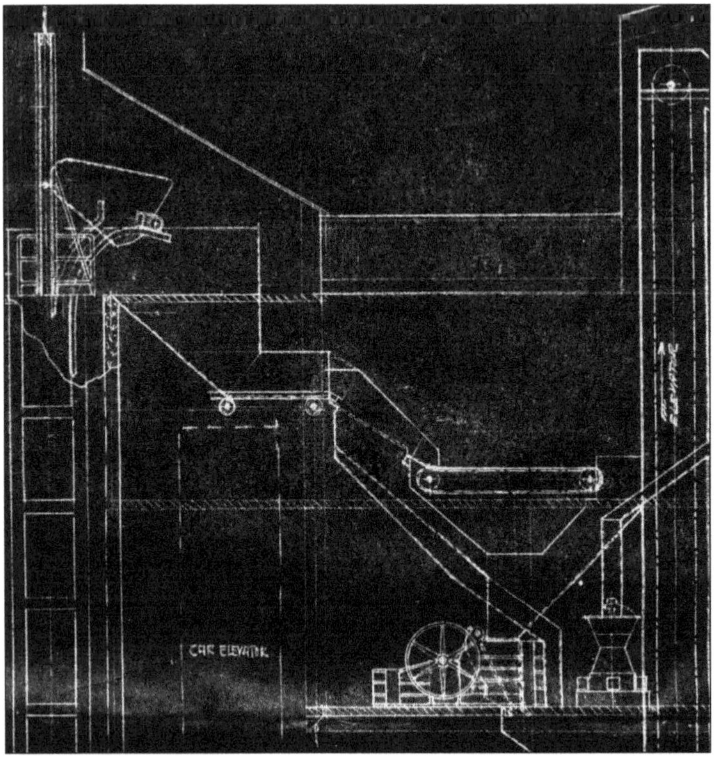

Once it was raised topside, the salt emptied into a six-inch screen. Smaller salt would fall through the screen and bypass a large crusher. Before 1951, large chunks, or lump salt, would be directed to a motorized picking table and then to the first large crusher. On this 1922 blueprint, the skip is shown on the left, tilted as if emptying the salt. The long oval represents the picking table. (Courtesy Hutchinson Salt Company.)

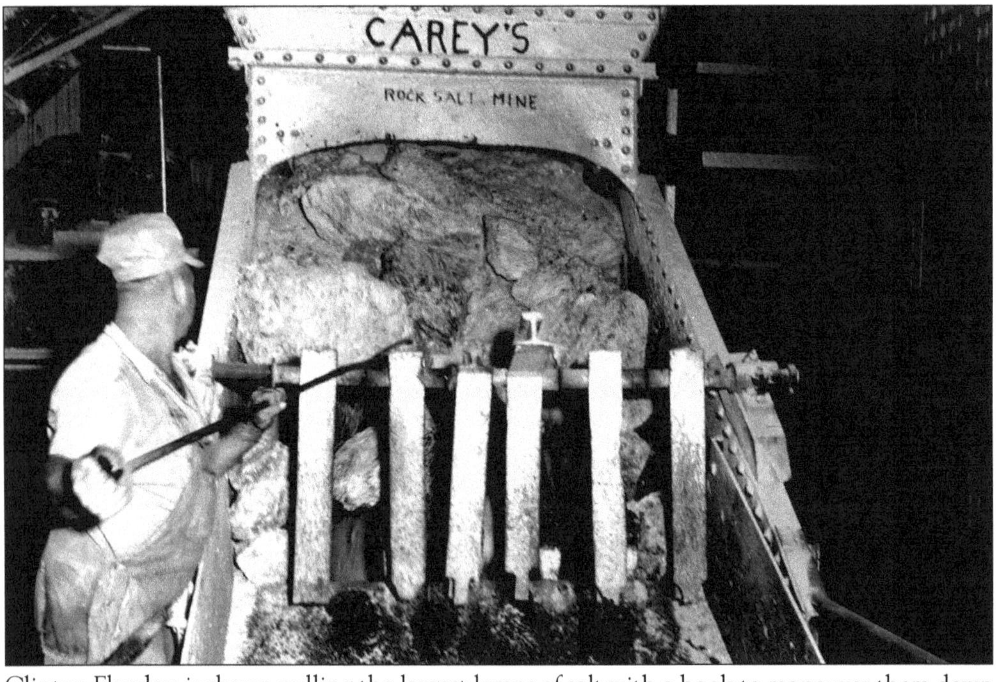

Clinton Flanders is shown pulling the largest lumps of salt with a hook to maneuver them down to what used to be the picking table. After the picking table was removed, the lumps merely slid down this ramp into the jaw crusher for the first reduction in size. This screen, with its six-inch bars, was nicknamed the "Grizzly."

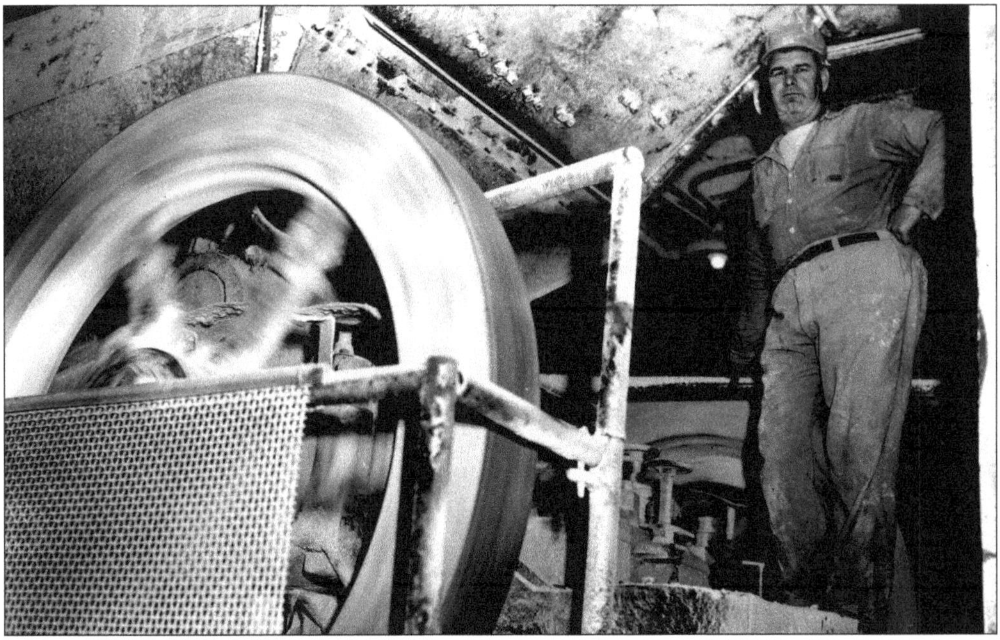

The jaw crusher is shown here, again with crusher operator Clinton Flanders. The Allen-Garcia Company considered the jaw crusher to be the most important piece of mill equipment because the rest of the mill operations would not to function properly if this equipment did not do its job.

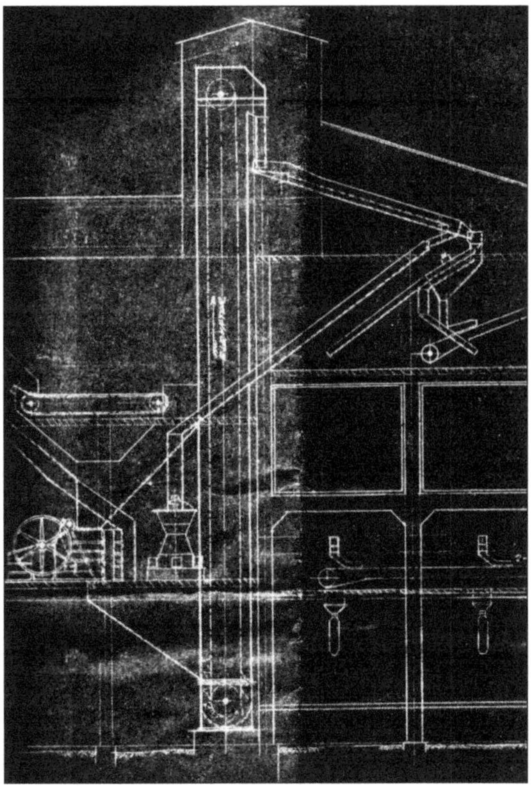

The reduced salt then traveled on the conveyor to an elevator, which carried the salt to the top of the mill to begin its journey to the screens.

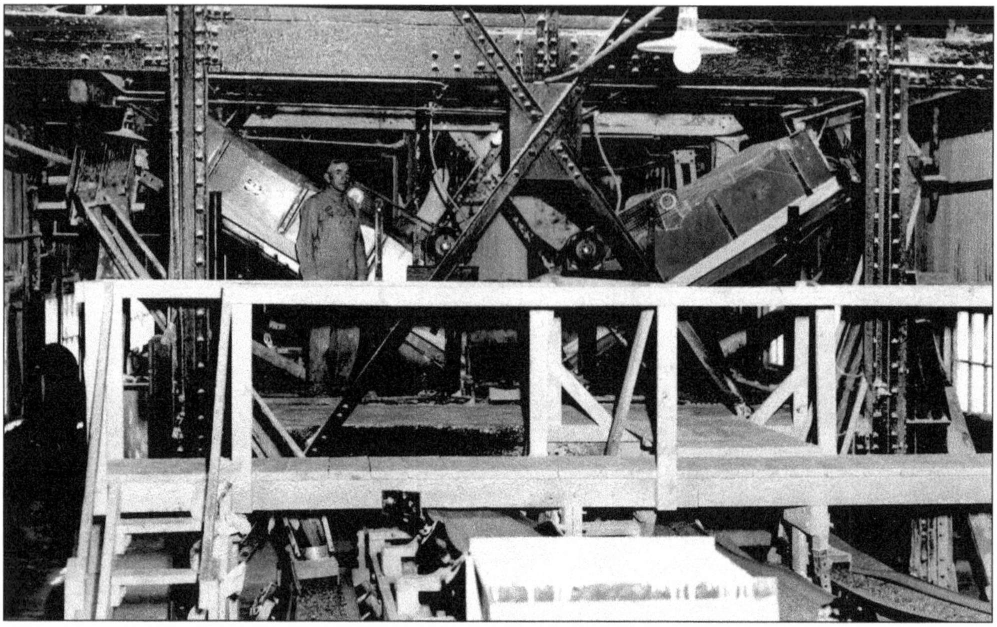

Three screens of progressively smaller mesh separated the salt into the different grades. Gravity helped the salt travel from one screen to the next screen beneath it. This 1952 photograph shows screen operator Claude Ryan standing next to the lowest bank of screens. The middle bank can just be seen below the pitched truss toward the top. Below Ryan's feet, the separated salt is moving to the storage bins on conveyors.

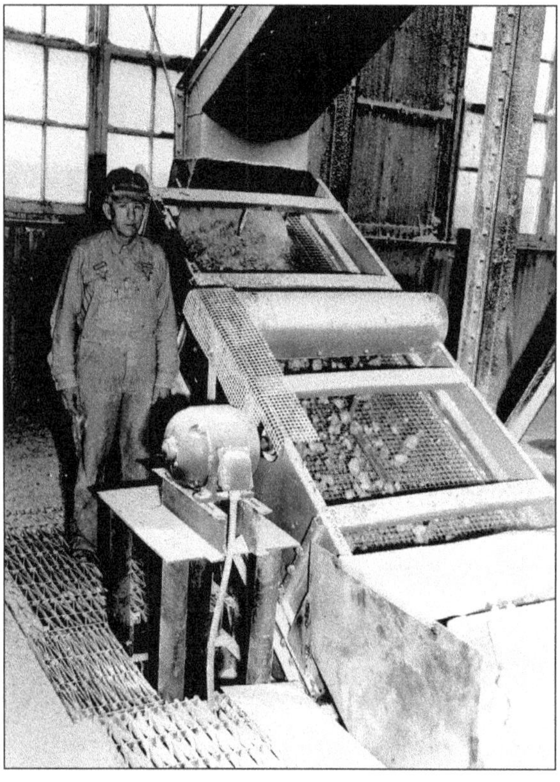

The screens vibrated, shaking the smaller pieces of salt through the mesh. Claude Ryan stands next to one of the screens in this photograph. The screen's motor sits in front of him. Salt too big to fit through the second and third banks of screens were either directed to yet another crusher or to the storage bins, depending on the demand for each varying grade of salt.

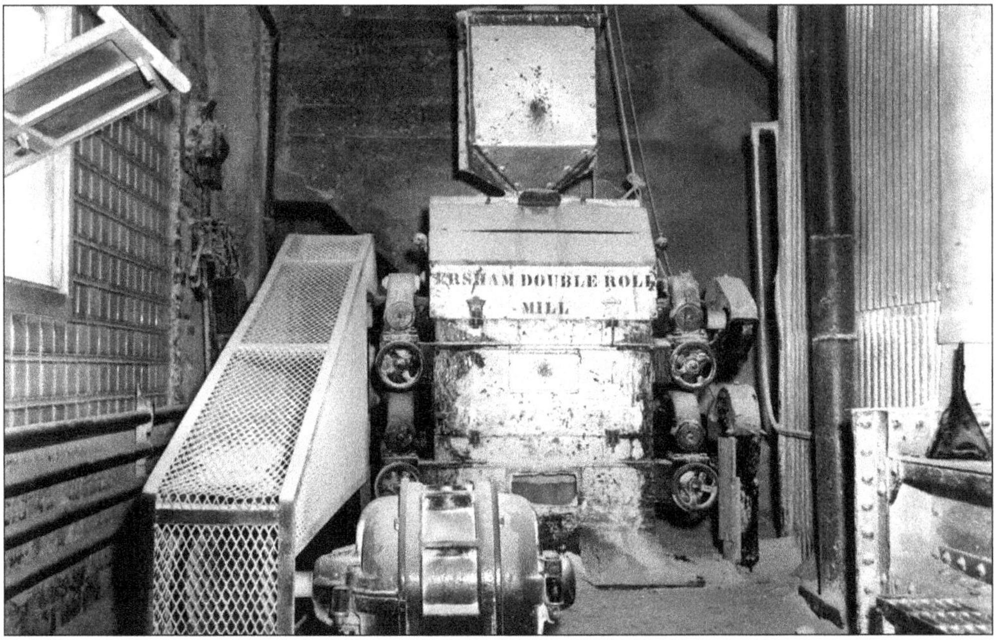

The Erhsam Company (whose name is misspelled in this photograph) built the two crushers that, if needed, would grind the salt into the finest two grades. These double-roll mills were commonly used in the other Hutchinson-area salt plants and cost the Carey Salt Company $2,150 for both in 1923. The Erhsam Company was headquartered in Hope.

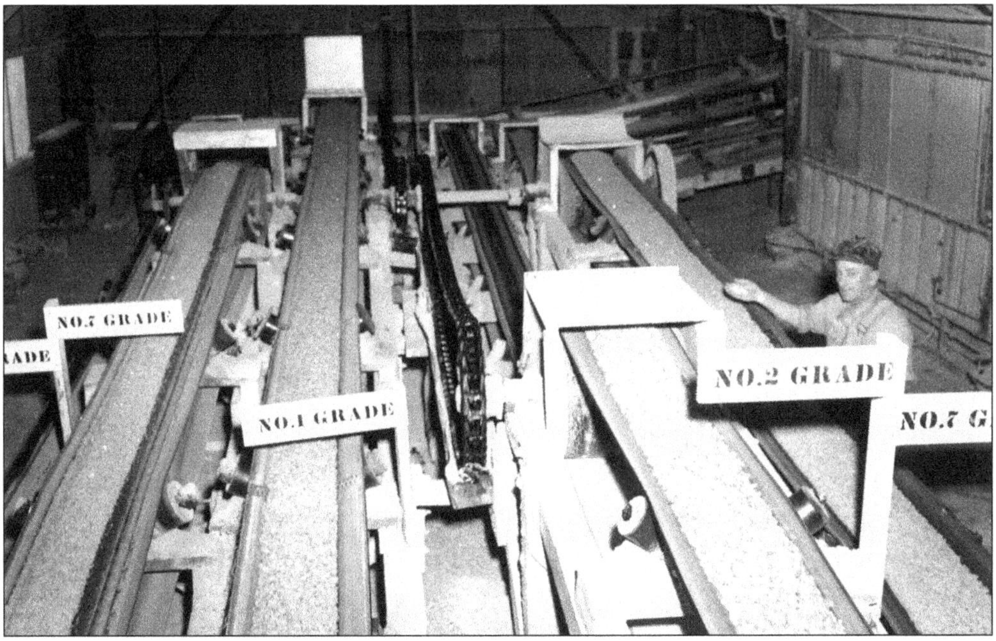

Since 1923, the mine had produced four grades of salt numbered 1, 2, 4, and 7. This 1955 photograph shows the four grades of salt traveling on the conveyor belts in the mill. A board of directors report stated, "The nomenclature . . . was most confusing to new customers, as well as to the public in general." In 1958, the Carey Salt Company switched to three grades named coarse, medium, and fine.

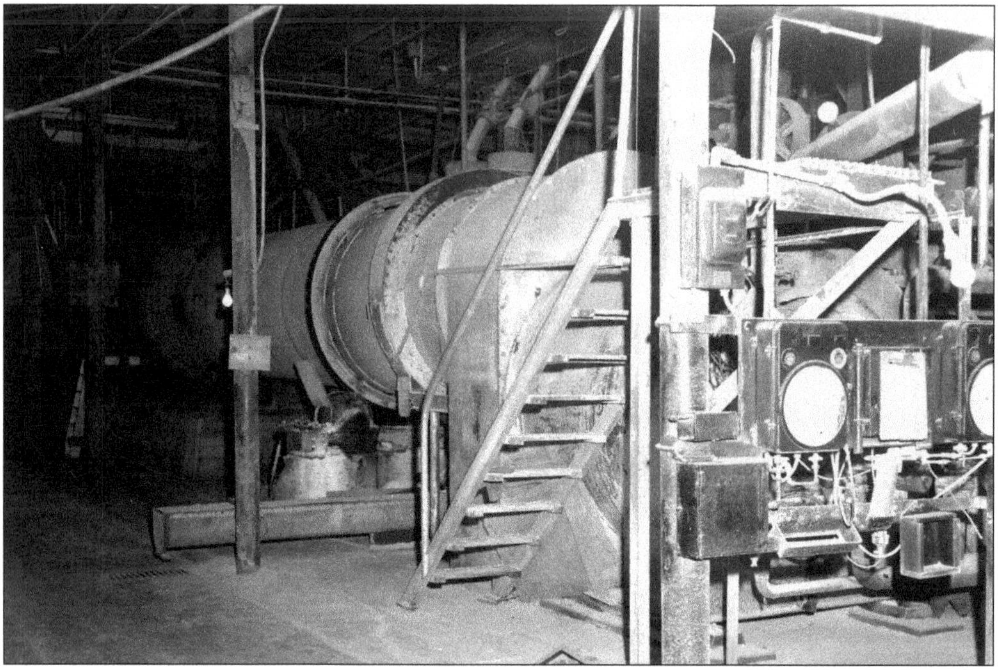

The Carey Salt Mine installed a rotary kiln, similar to this one from the evaporation plant, in the 1980s to satisfy the needs of a client making animal feed. Salt fines were placed in the kiln, which rotated. The heat from the kiln dried the fines to keep it from caking. This salt often had minerals added to it and was commonly mixed with livestock feed.

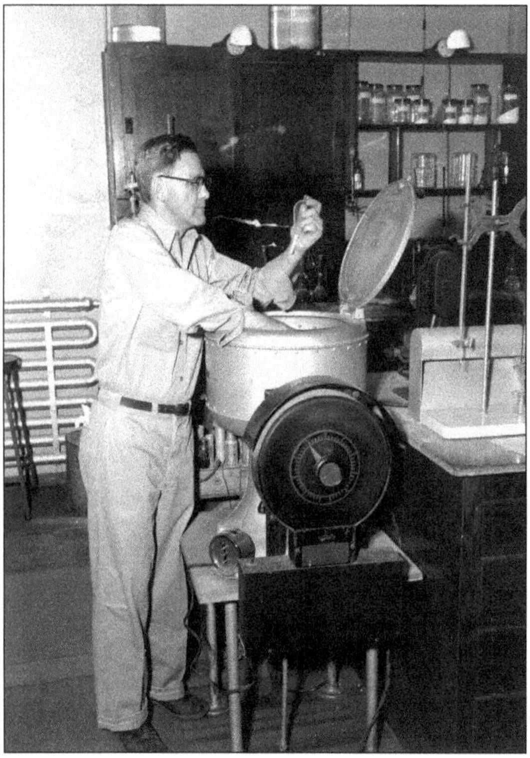

The Carey Salt Company's laboratories, opened in 1925, were shared by the evaporation plant and the mine. The chemists tested the rock salt for impurities and advised the company on which products to add to the salt. As the salt industry became more complex, the chemistry department also grew, developing new products and ensuring that the existing ones were up to par. This 1952 photograph shows control chemist Phil Wharton testing a sample.

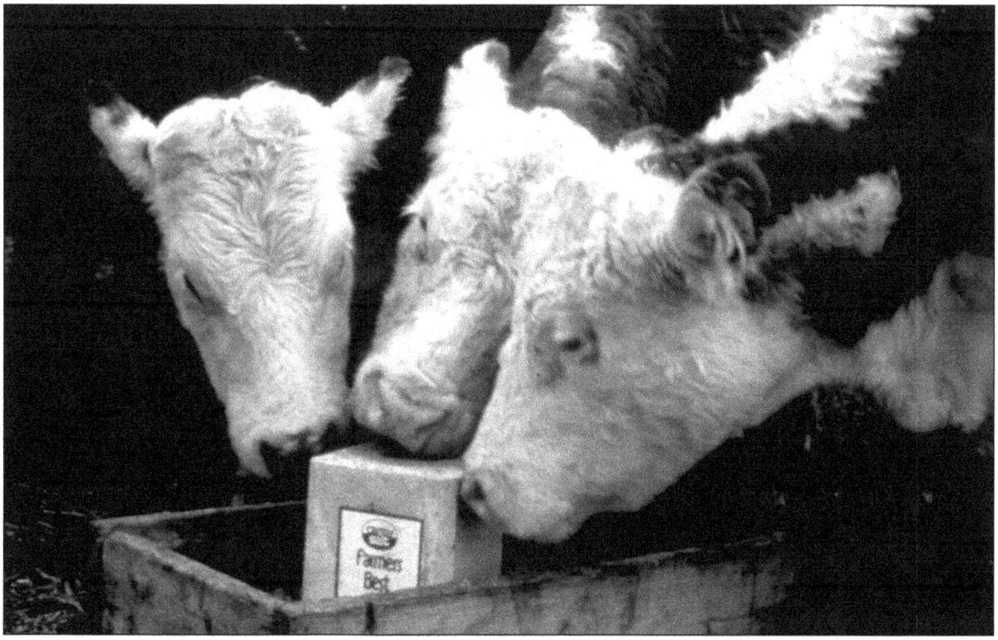

Although original plans included two block presses in the mill, they were never installed. Instead fines from the mine were shipped to the evaporating plant and pressed into a gray block to add to Carey Salt's line of mineral blocks. For many years, farmers could also purchase lump salt to place in their fields. Blocks were preferred because, although the lump salt was inexpensive, the salt often cut the animals' tongues.

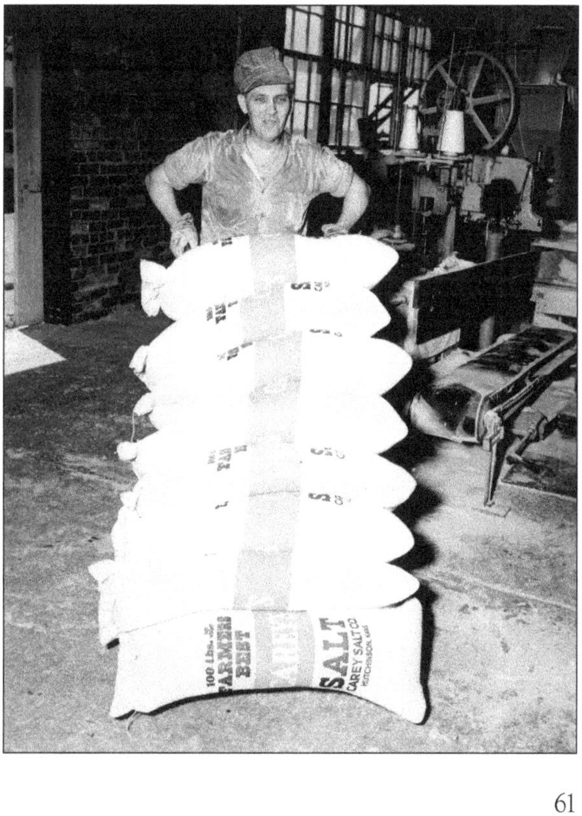

Bob Neill loads sacks of Farmers Best salt, Carey's earliest rock salt product, onto a railcar. The sacking and stitching machine can be seen in the background. Carey Salt used cloth bags since it began its operation in the 1920s. Consumers liked the cotton sacks because the bleached cloth was useful for making towels and other items. Paper sacks began replacing the cloth ones in the 1950s.

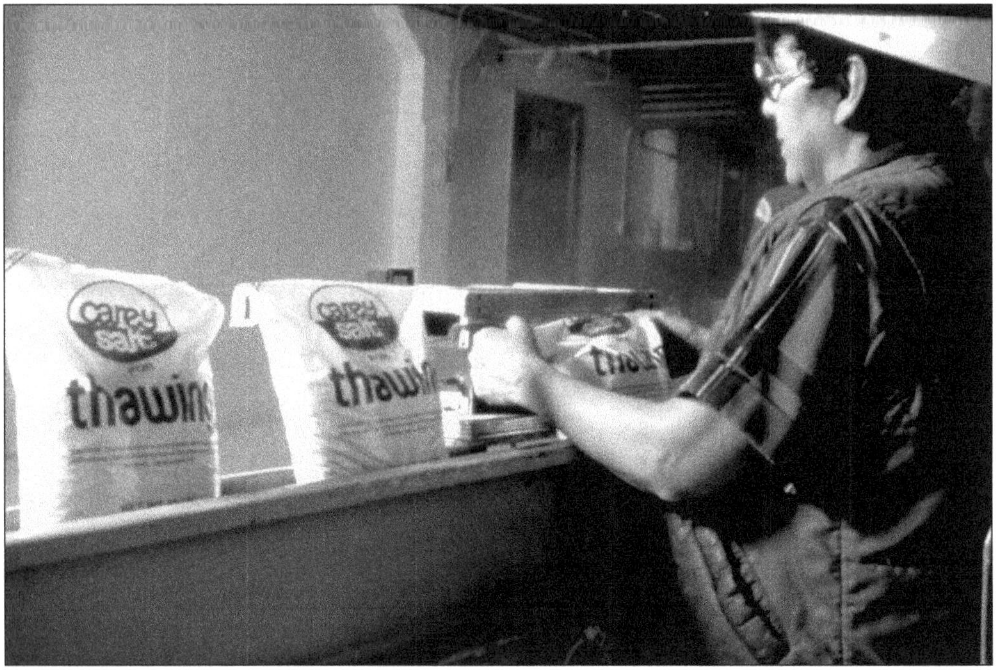

Finally plastic sacks were introduced. Workers closed the paper and cloth sacks by means of an industrial sewing machine, but the plastic sacks were sealed by heat. Instead of duplicating the process at the mine mill, the rock salt was sent to the evaporation plant, which already had the equipment, for packaging. This unidentified woman is sealing sacks of consumer-grade thawing salt in this 1980s photograph.

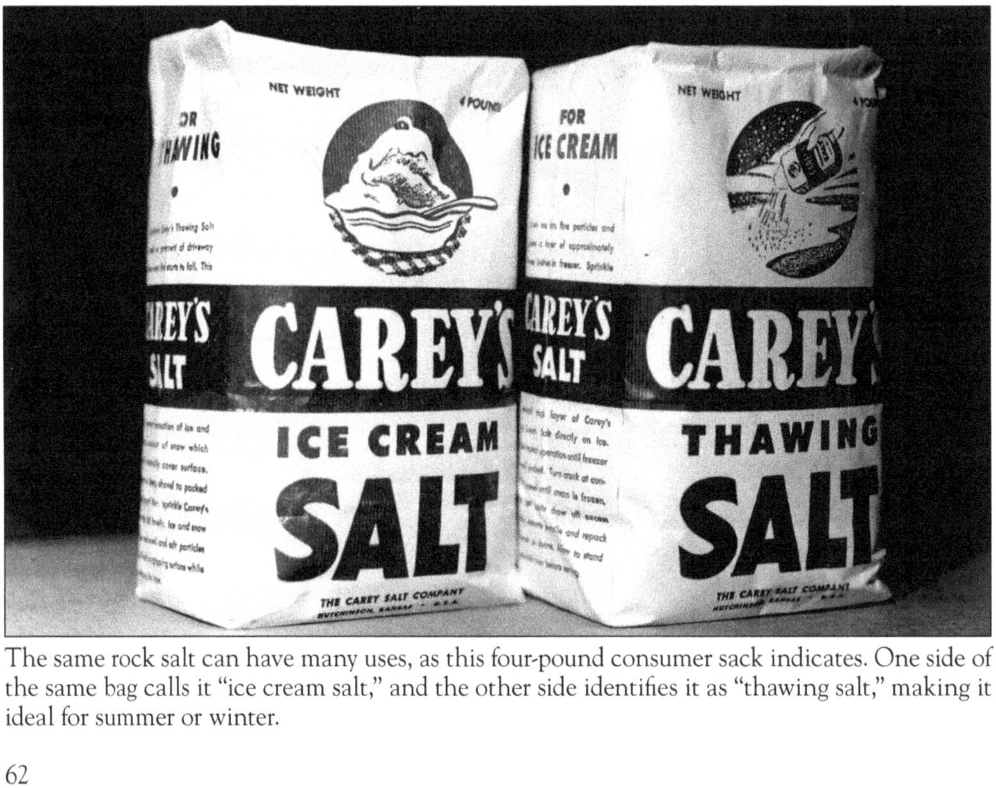

The same rock salt can have many uses, as this four-pound consumer sack indicates. One side of the same bag calls it "ice cream salt," and the other side identifies it as "thawing salt," making it ideal for summer or winter.

The coarse salt was sold to various industries, including companies that tanned hides. Jake Alexander sprinkles salt over several hides at the Winchester Packing Company in Hutchinson in this photograph. In this method, the dry, coarse salt is spread over the wet hides to remove the moisture within the skins. This cures the hides, stemming decay, before tanners can begin the process of making leather.

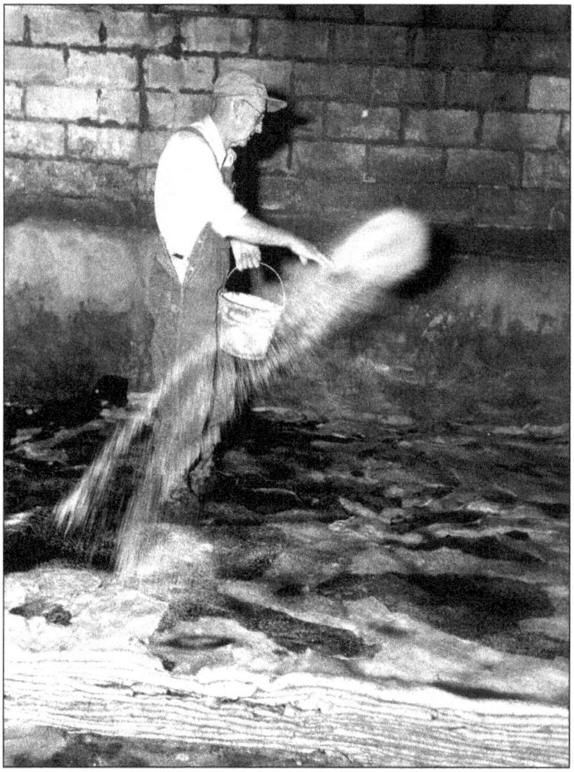

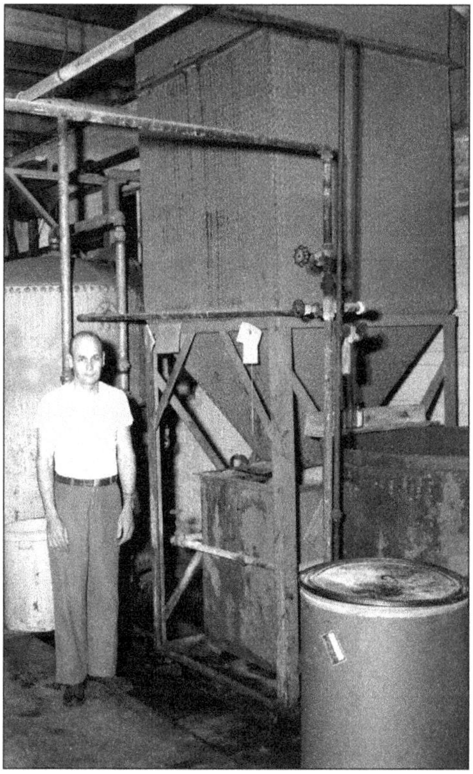

Sales manager Paul Toland discovered a need for a machine that turned rock salt into brine. With Jack Hinrichs, a former technical engineer at the Carey Salt Company, he invented the Brine-O-Lator. Carey marketed this patented machine, which was built by the C. E. Johnson Sheet Metal Company in Hutchinson, to commercial laundries and other industries nationwide. In this photograph, J. T. Dunsworth of Boone Ideal Laundry in Hutchinson poses next to his Brine-O-Lator.

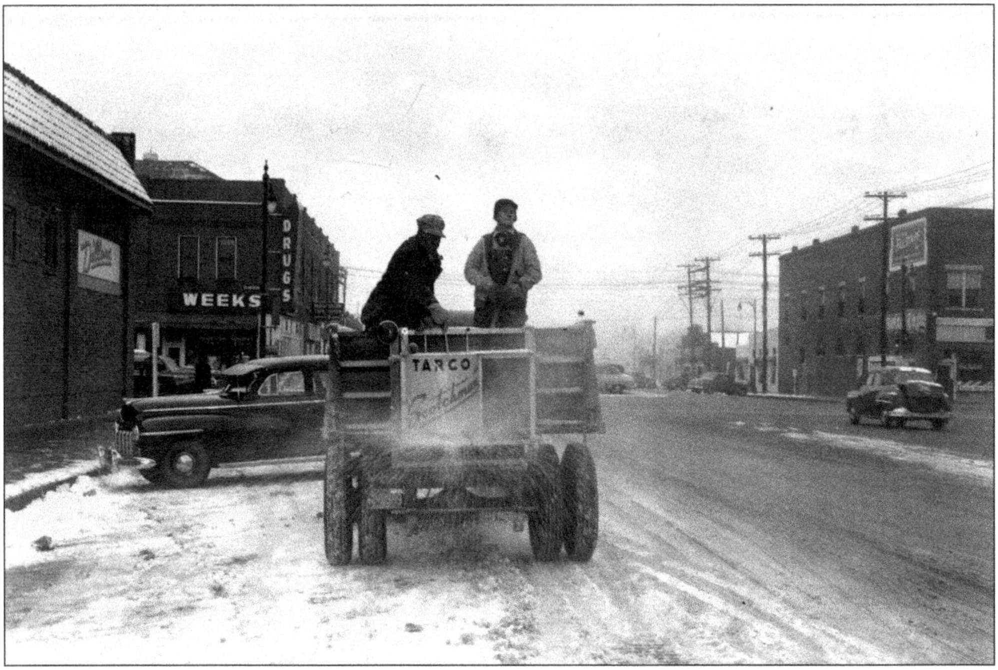

In the 1950s, problems with corrosion became controversial in the use of salt on roads. The Carey Salt Company began mixing salt with Banox, a rust inhibitor developed during World War II. The use of rock salt for deicing roads has grown to become the bread and butter of the rock salt industry. This 1952 photograph shows Hutchinson city street workers sprinkling Banox-treated salt on Avenue B, near Main Street.

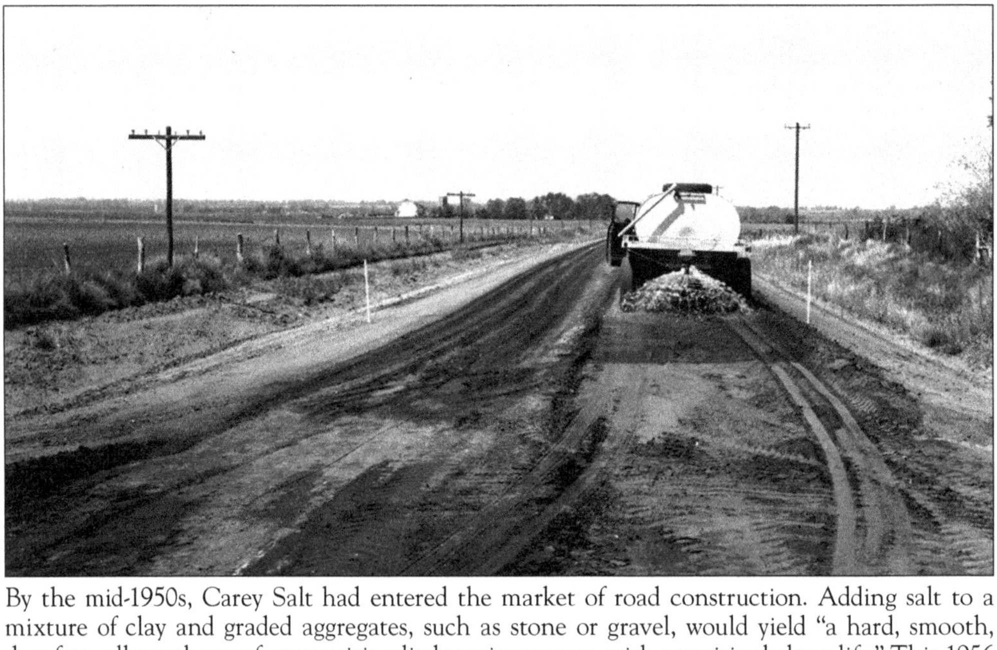
By the mid-1950s, Carey Salt had entered the market of road construction. Adding salt to a mixture of clay and graded aggregates, such as stone or gravel, would yield "a hard, smooth, dust-free, all weather surface requiring little maintenance, with surprisingly long life." This 1956 photograph of a rural Reno County road was published in a booklet illustrating the steps of road stabilization.

Six

GETTING THE SALT TO THE CONSUMER

Once the salt was processed, distribution to the markets was key. The salt trusts during the early years of the Carey Salt Company forced the small, independent companies to battle for the right to send their salt on the railroads. The decision of the Carey Salt Company's executives to open the Winnfield and Cote Blanche mines in Louisiana had everything to do with the railroad freight rates. The Kansas mines could only send salt profitably to certain states, none of which were in the southern market. If Carey Salt had learned anything over the years, it was that with poor distribution, one might as well close up shop. This attitude affected the company on the level of navigating through markets to properly loading the railcars on the loading dock.

In August 1923, Gov. Johnathan Davis had barely returned to Topeka after speaking at the mine's opening dedication. During this same month, the Kansas Public Utilities Commission authorized Emerson Carey's new Hutchinson and Northern Railway (H&N) line to "be operated as a terminal railway and charge fees for switching services." This new line, which in 1926 would become the shortest line eligible to conduct interstate commerce in the country, would be a vital component of Carey's salt operations because it would connect the two plants to the main lines of not only the ATSF but also the Chicago, Rock Island and Pacific and the Missouri Pacific Railroads running through Hutchinson. The line operated throughout Carey Salt's history.

The Allen-Garcia Company built sophisticated methods of loading railcars into the mine's architecture in 1922. Experience in shipping products from the evaporation plant had taught the Careys the importance of successful delivery, and steps were taken to ensure this process went smoothly at the mine.

This philosophy was the same with the trucking industry. Carey Salt had used trucks since the 1920s, but shipping on rails far surpassed trucking until the 1950s, when trucking began to steadily gain. In 1944, Carey Salt decided that trucking was so important an entirely new wing was added to the mill building simply for this purpose.

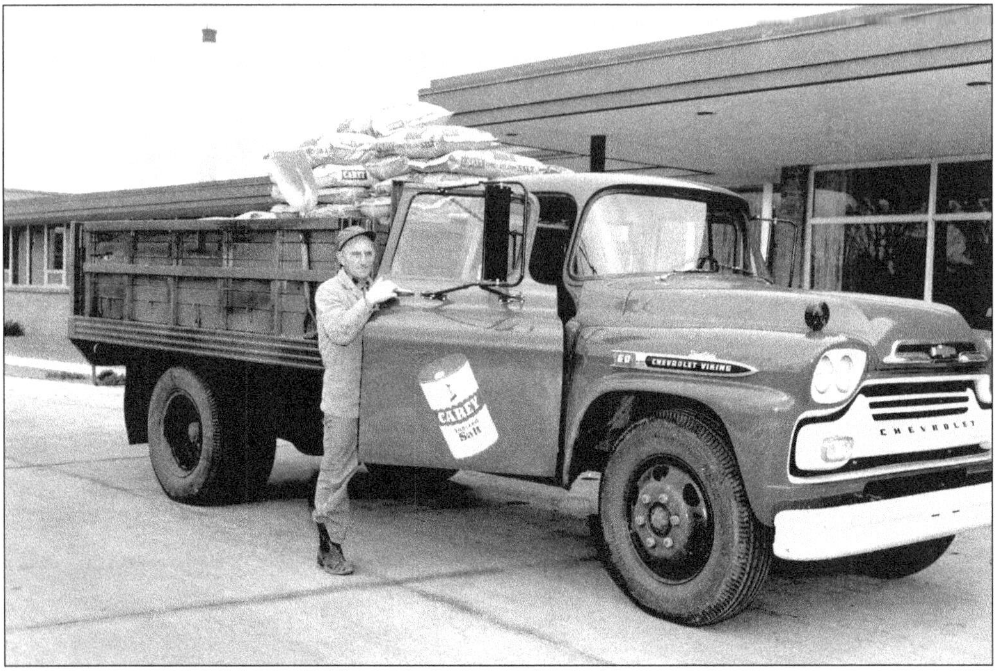

The Carey Salt Company employed a smaller truck to make deliveries around Hutchinson. Ernest Hippen, shown here in 1959 in front of the corporate offices with his new Chevrolet Viking, drove bags of rock salt to local businesses, including the Winchester Packing Company, J. S. Dillons, and Western Foods.

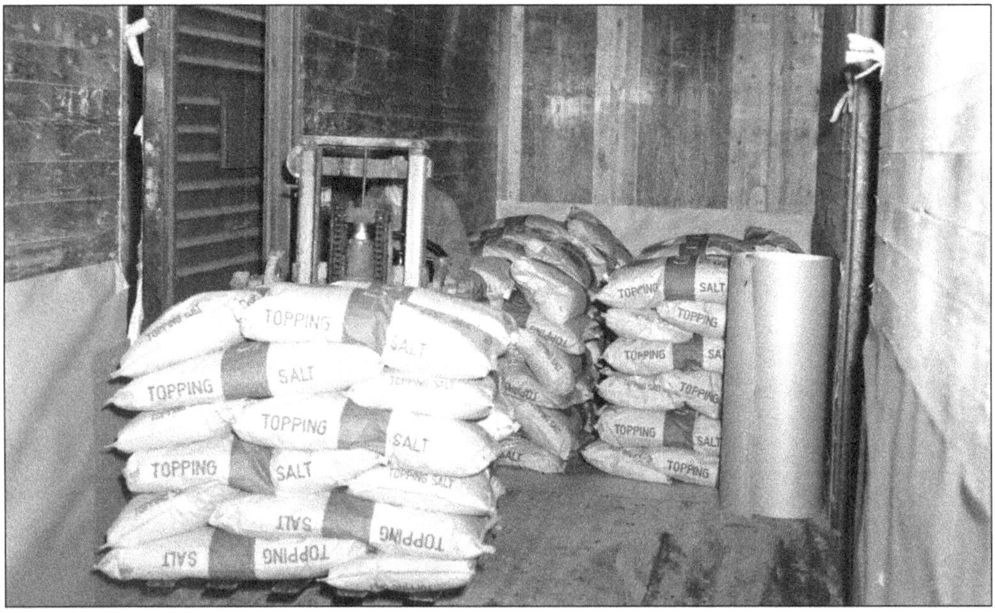

A 1925 testimonial stated, "Have just unloaded my car of salt and it goes without saying that the pride taken in loading a car of various varieties of Carey Salt is manifested in opening the car and gazing into it." Ensuring the salt arrived unsullied and unbroken was a priority for Carey Salt. Loaders lined the boxcar with paper to keep the bags clean and to prevent ripping from protruding nails.

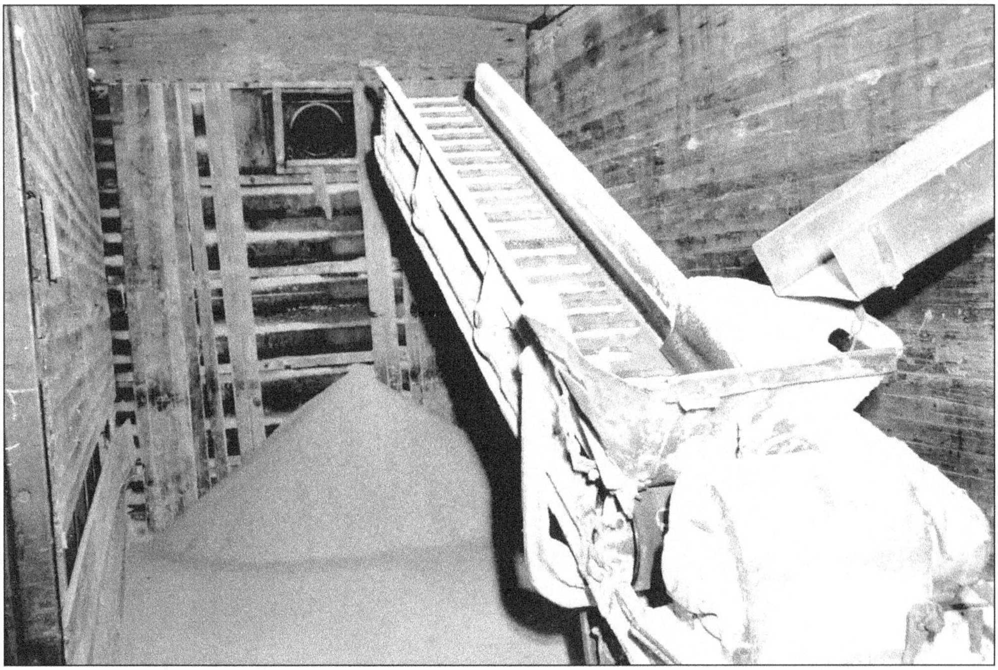

As early as the mine's opening day, the mill had a loading chute made by the Manierre Engineering and Machinery Company that could automatically fill a railcar, as shown on the right of this photograph. By the 1950s, the boxcar loaders were using this flight elevator to fill all areas of the car.

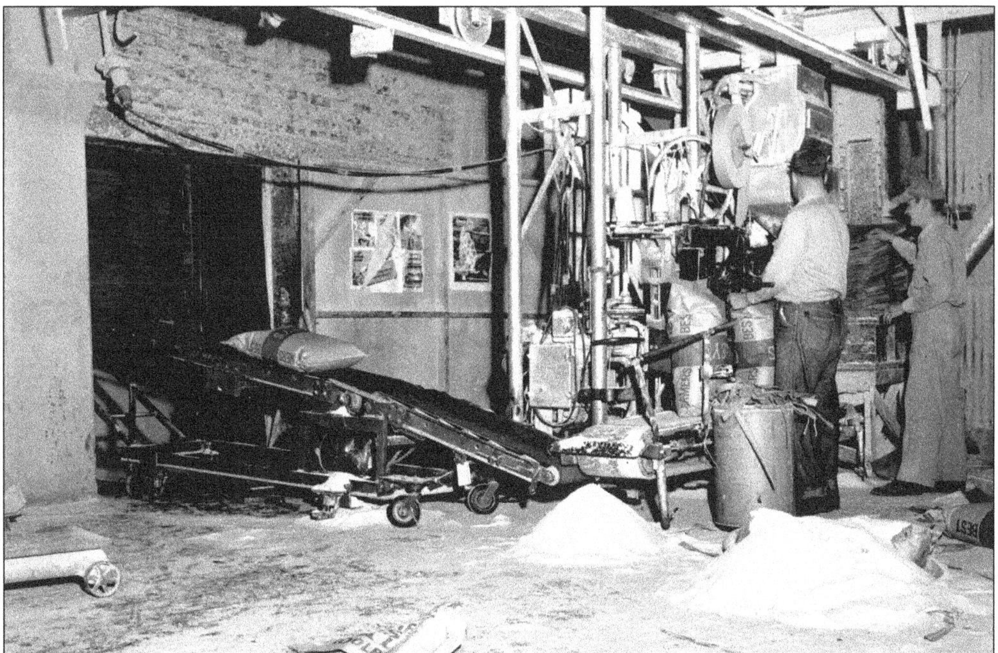

The rock is placed into bags at the mine mill in 1953 in this photograph. Bud Daniels (right) fills the sack upside down, and Lloyd Musick operates the sewing machine to stitch the bottom of the sack before it is loaded directly into a waiting boxcar.

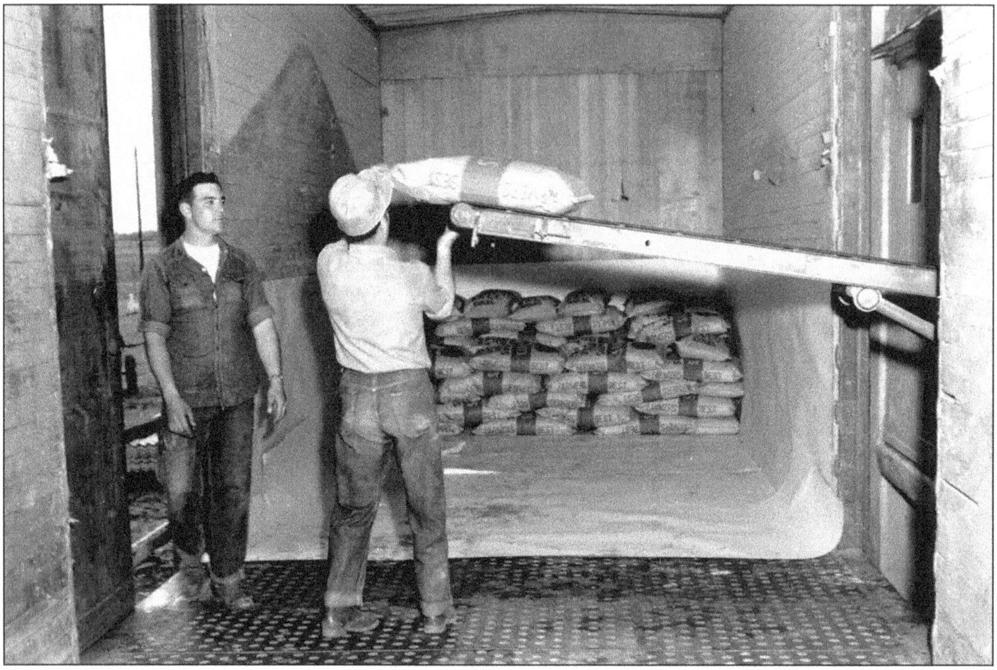

The new loading conveyor was featured in the June 1953 issue of *Salt and Pep*. "Inside the car Tommy Delventhal, Dock Hand, takes [the sack] off the conveyor, while Jake Herdt, Storekeeper waits for another to come up the conveyor. Previously the sacks had to be trucked on two-wheel trucks."

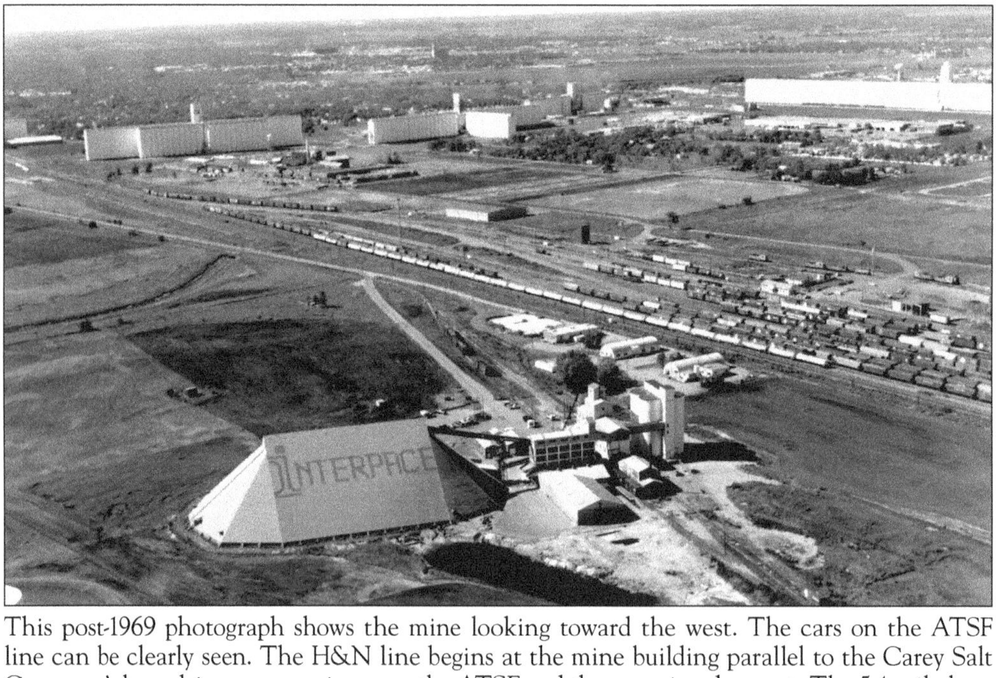

This post-1969 photograph shows the mine looking toward the west. The cars on the ATSF line can be clearly seen. The H&N line begins at the mine building parallel to the Carey Salt Company's long driveway, running near the ATSF and then veering due west. The 5.4-mile-long railway traveled along the Carey Salt evaporating plant as well, picking up items for transfer to the major railroads.

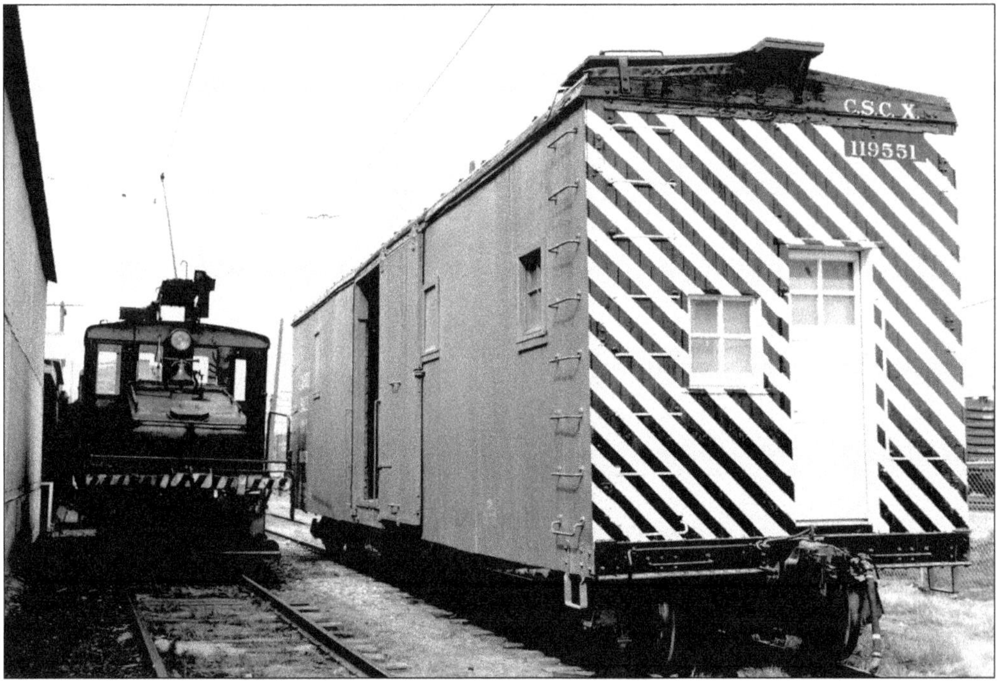

Transport of salt and supplies between the evaporation plant and the mine were common. The Carey Salt Company purchased *Lulubelle*, a freight car designated for transport between the two facilities, in the 1950s to run on the H&N line.

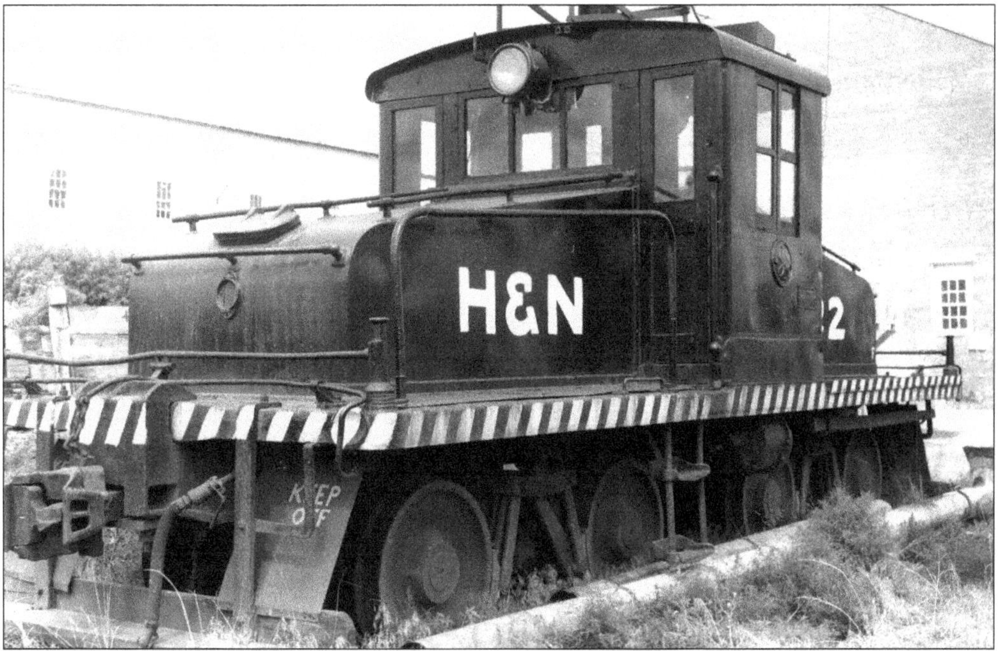

The first H&N engine, a 30-ton General Electric, was purchased new in 1923 when the line opened. In 1926, the H&N bought No. 2, the 1919 General Electric locomotive shown here. Midwest Iron and Metal of Hutchinson purchased the train in 1963 and eventually donated it to the City of Hutchinson, which in turn gave it to the Reno County Historical Society in 2000.

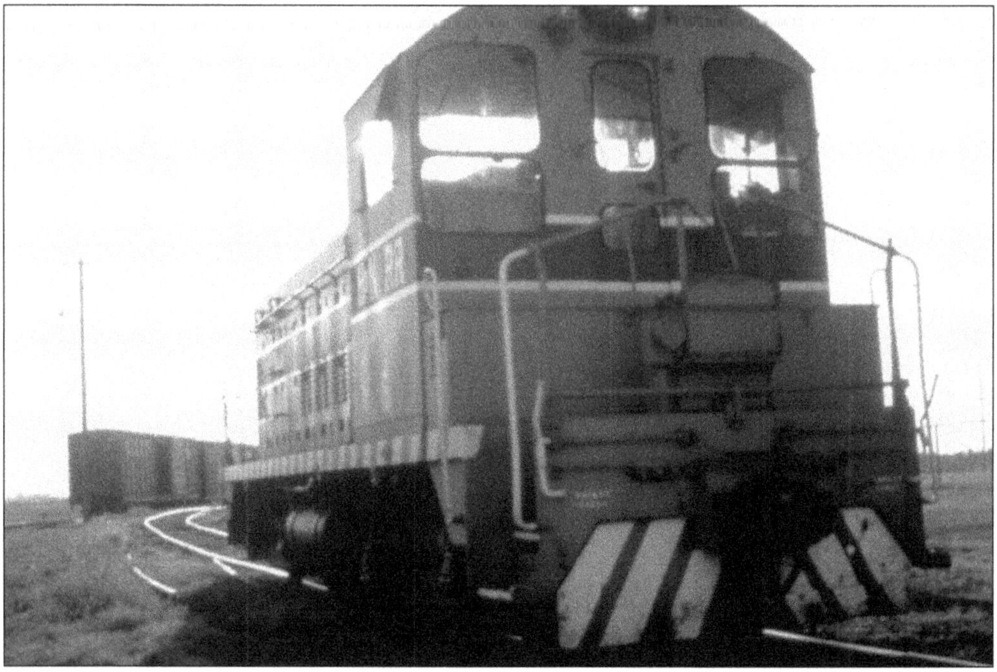

Electricity from the Carey Salt Company evaporating plant operated the trolley lines until 1969, when the H&N finally switched to diesel power. Locomotives No. 4 and No. 5 were identical 100-ton, diesel-electric engines with 660 horsepower. Pictured here is locomotive No. 6, a 900-horsepower engine built in 1957 by Electro Motive Division of General Motors. In the 1980s, the engine was operated along the H&N by David Smiley and Darrell Renner.

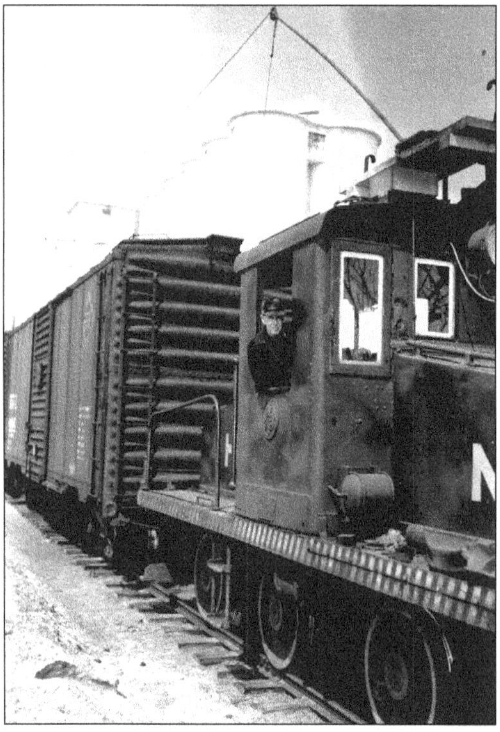

The H&N provided lucrative switching services to the industries on the eastern side of town. For example, the neighboring Champlin Oil Refinery was paying over $500 per month in 1956 to the H&N for hauling tank cars to the main lines. This photograph shows engineer Harry Stephens driving engine No. 1 past the Kelly Flour Mill. A former streetcar engineer, Stephens drove for the H&N from 1925 until 1951.

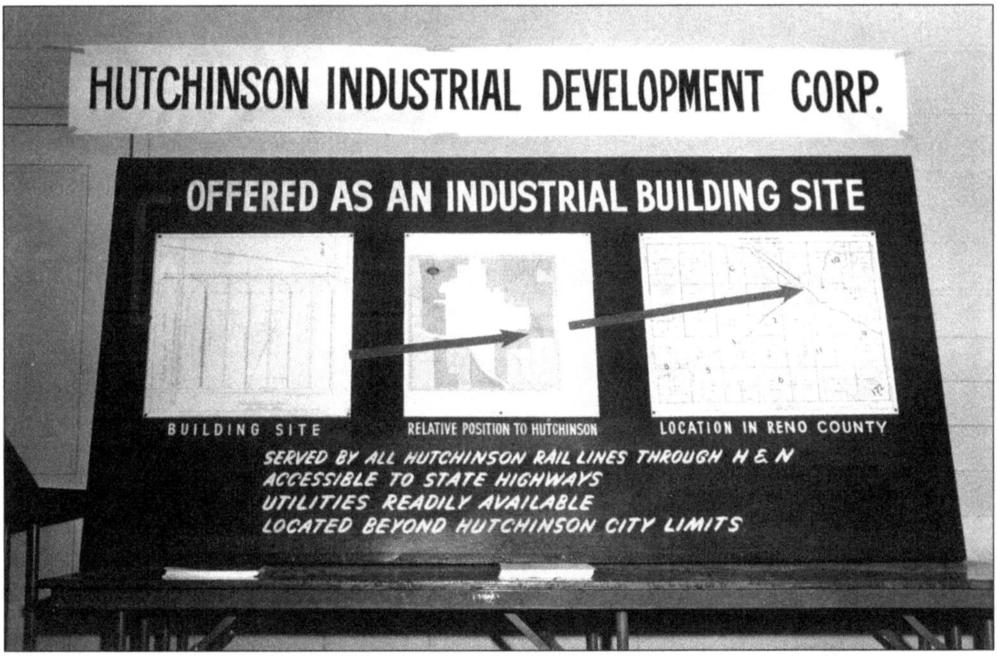

This display sat on a table at an industrial show in the Hutchinson Sports Arena in 1956 to entice businessmen to locate their industry on Carey grounds. The land was located on the H&N line and would bring the Carey company revenue via the H&N's switching services.

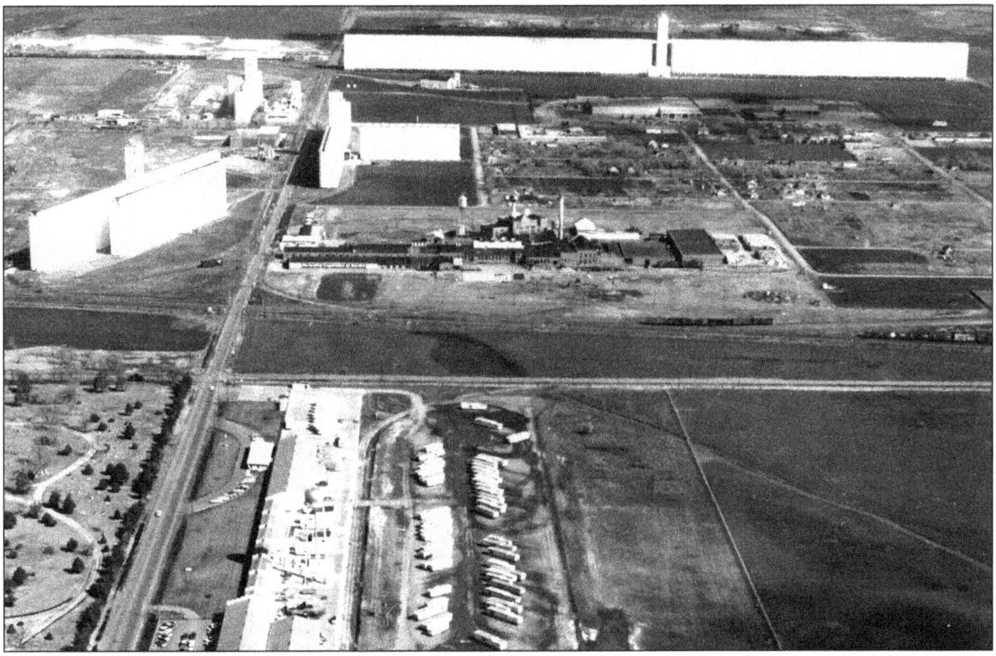

Detroiter Mobile Homes submitted plans to develop a facility in the Hutchinson Industrial Development Corporation's district in 1958. Carey executives estimated that nearly 400 cars would travel to and from the plant annually, with a switching service fee at $12.80 per car for the H&N. This 1964 photograph shows Detroiter Mobile Homes in the foreground. In the middle is the old Central Fibre Products plant, another Carey industry.

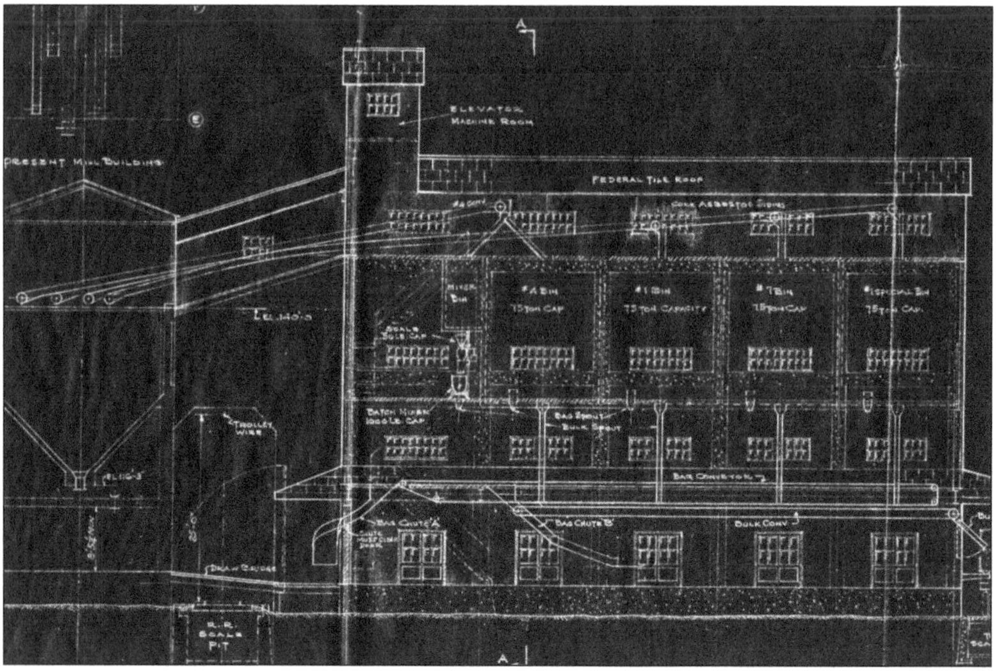

Trucks had been used to transport salt since the 1920s. Companies would bring livestock to town, and Carey Salt would fill the empty trucks with salt on their return trips. Although it offered a hefty cost savings, the unreliable drivers were a significant drawback. New regulations in 1935 improved the trucking industry, and in 1944, Carey Salt built a new building dedicated to handling truck usage. (Courtesy Hutchinson Salt Company.)

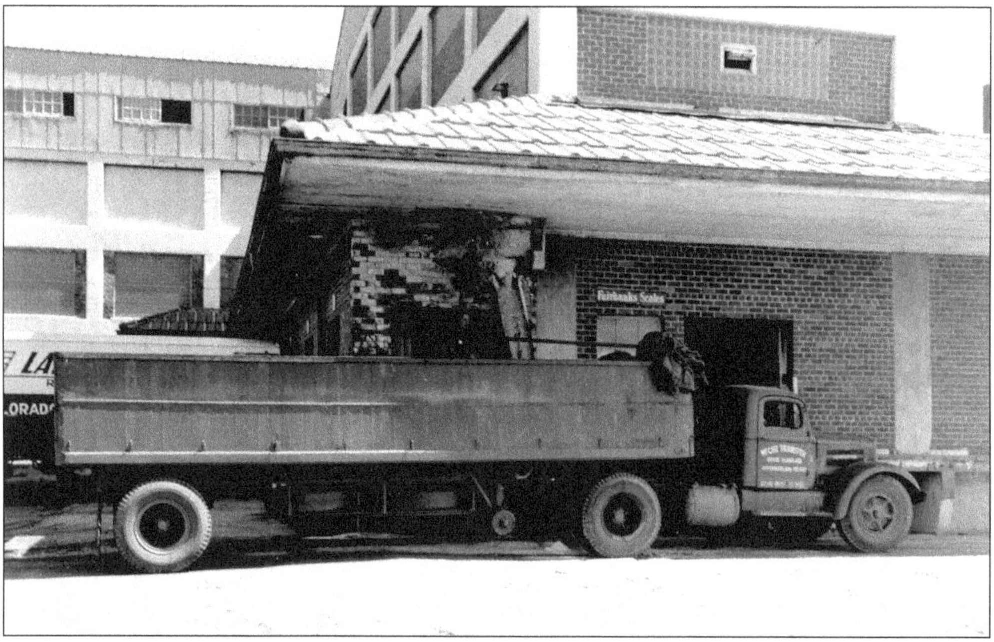

The truck in this 1951 photograph is loading at the bulk salt station on the end of the new building. The side of the truck says, "McCue Transfer Company, Grand Island Nebraska, Hutchinson Kansas."

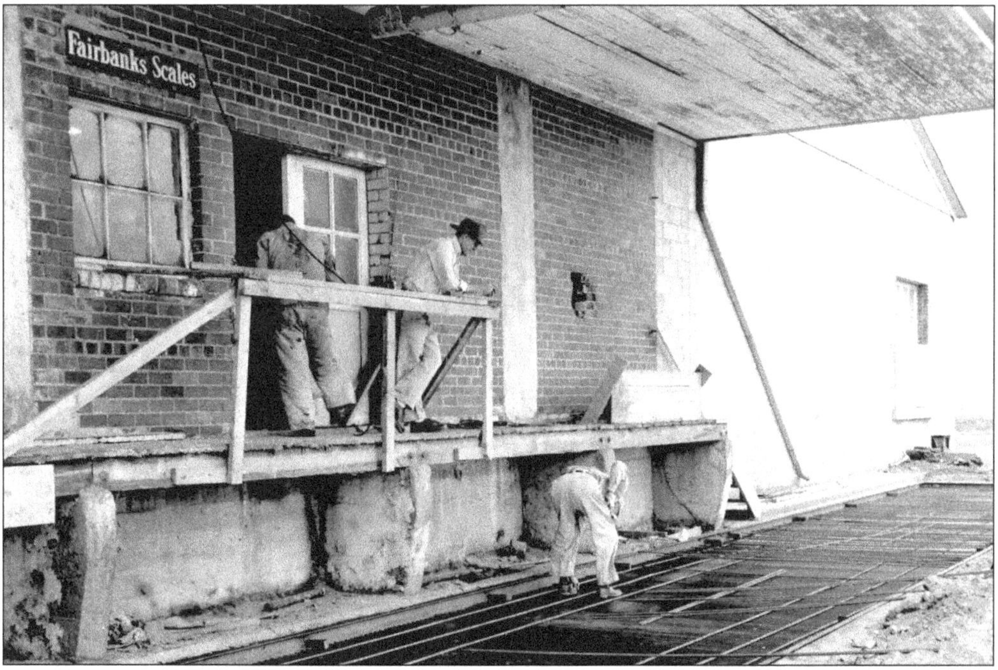

In July 1956, a new Kansas regulation went into effect, prompting the Carey Salt Company to install a new truck scale spanning 60 feet. Carey's practice of weighing a truckload in sections on its short 34-foot scale became illegal in this year.

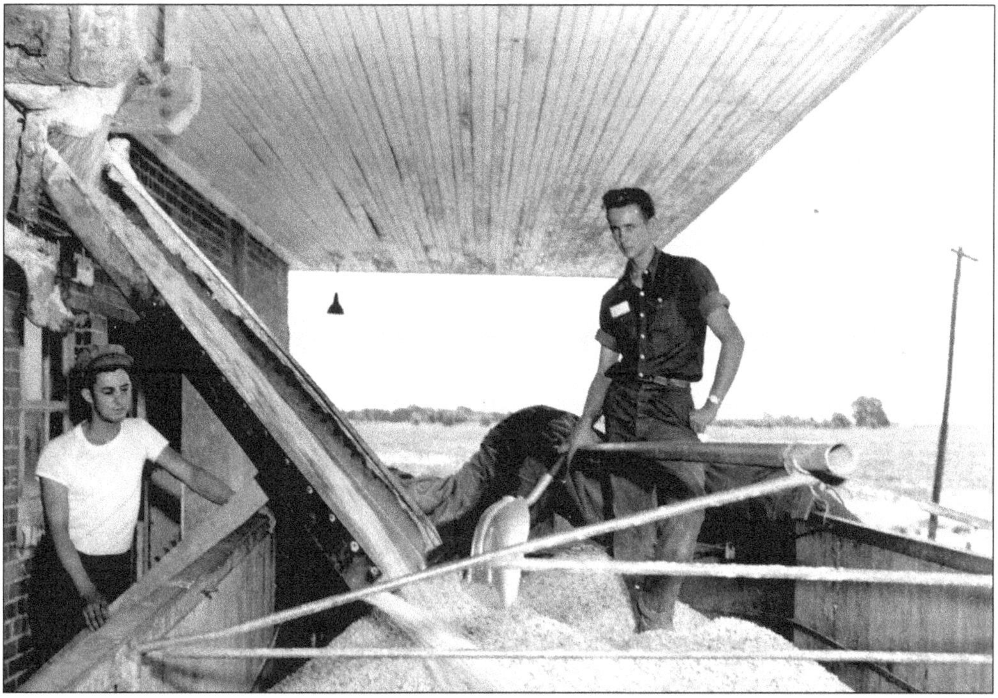

Carey Salt loader Wilbur Goheen (left) and an unidentified truck driver dump No. 2 salt into the truck bed. The driver stands ready with a shovel to distribute the load. Afterward a tarp would be stretched over the metal frame in the truck to protect the salt from the elements.

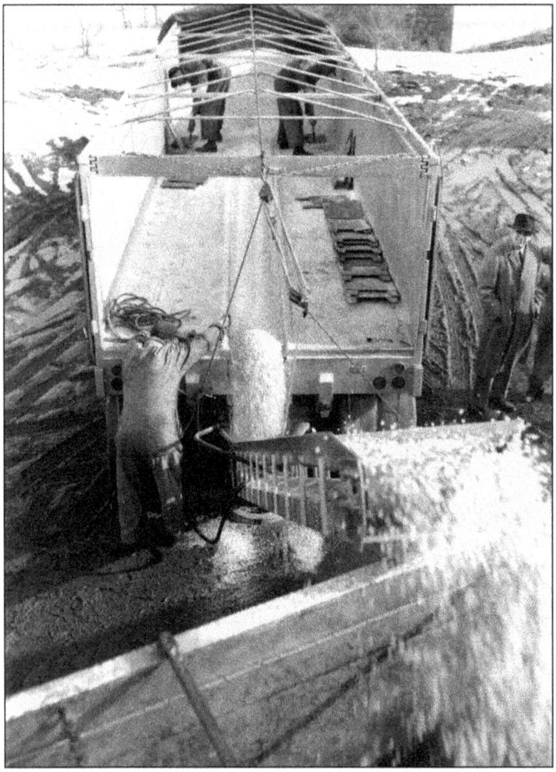

In 1958, the Carey Salt Company had a company design a new type of truck bed to facilitate unloading bulk salt. The bottom of the bed had a conveyor underneath removable plates, which dropped the salt onto an elevator to raise into a bin or other container. Workers were still necessary to shovel the salt into the conveyor. At the right, inside the truck bed, are the steel plates.

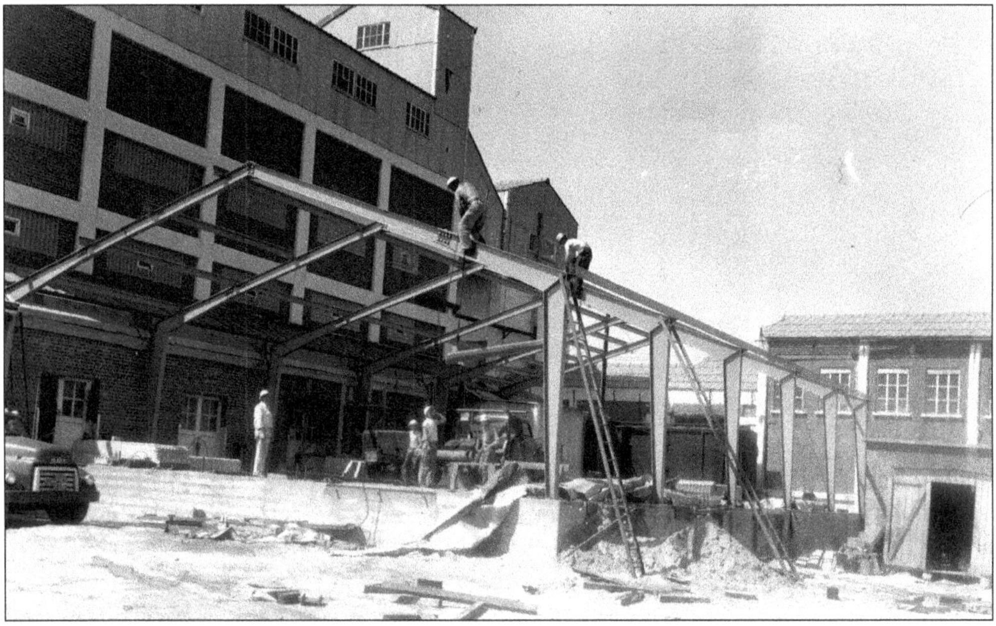

With the increasing demand for rock salt for use in ice control, and with big upgrades in capacity for production, there were times that the Carey Salt Mine could not keep up with demand. During off-peak times, it bagged as much rock salt as it could and stored it at the evaporation plant to await the winter storms. In May 1952, workers erected a building on mine property to house bagged salt.

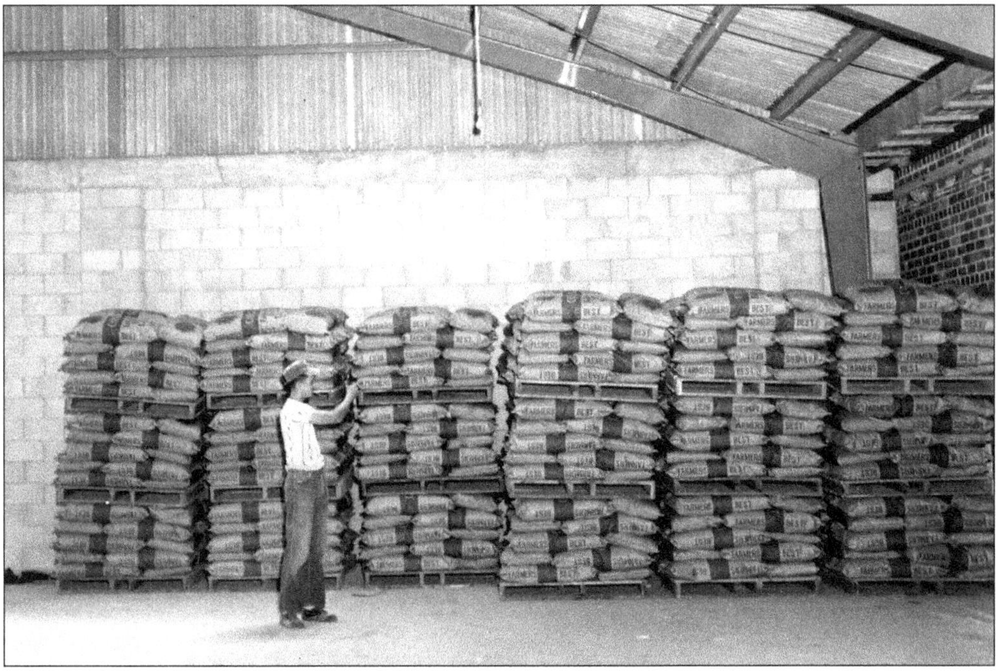

The new shed measured 50 feet by 83 feet and had a capacity of 20,300 cubic feet. As this photograph shows, it held bags of salt that were ready to ship. It was also intended to store empty sacks from Morton, Barton, and other salt companies. Carey Salt often packaged the salt under the labels of other companies.

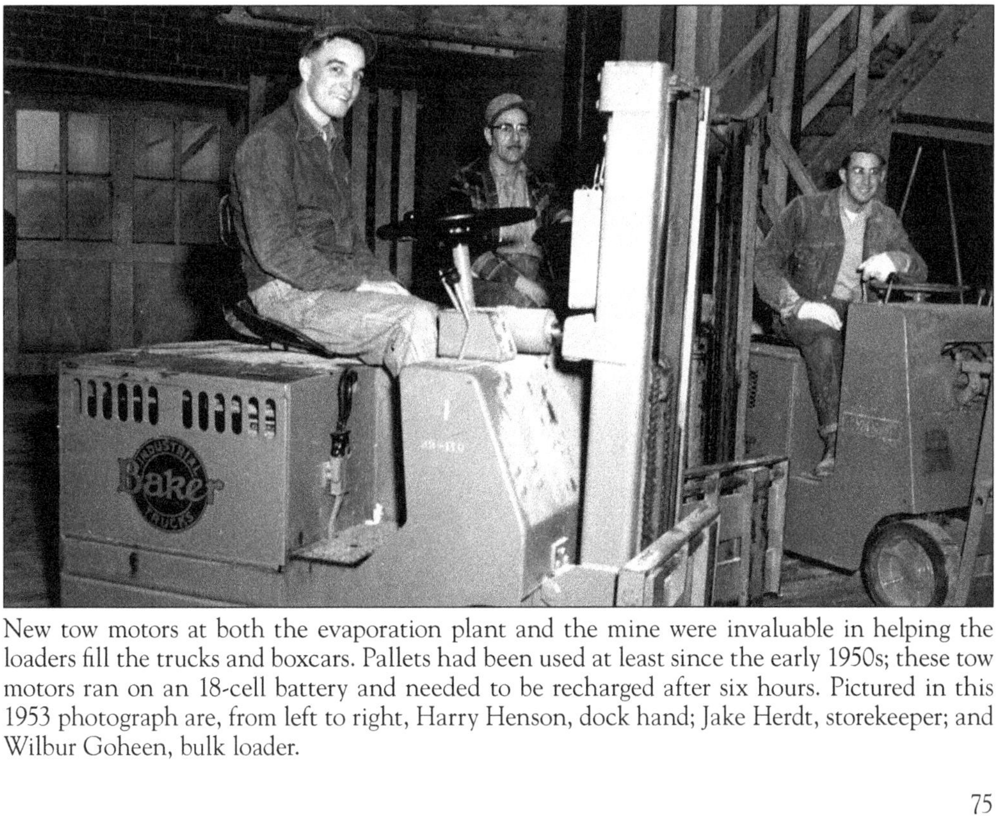

New tow motors at both the evaporation plant and the mine were invaluable in helping the loaders fill the trucks and boxcars. Pallets had been used at least since the early 1950s; these tow motors ran on an 18-cell battery and needed to be recharged after six hours. Pictured in this 1953 photograph are, from left to right, Harry Henson, dock hand; Jake Herdt, storekeeper; and Wilbur Goheen, bulk loader.

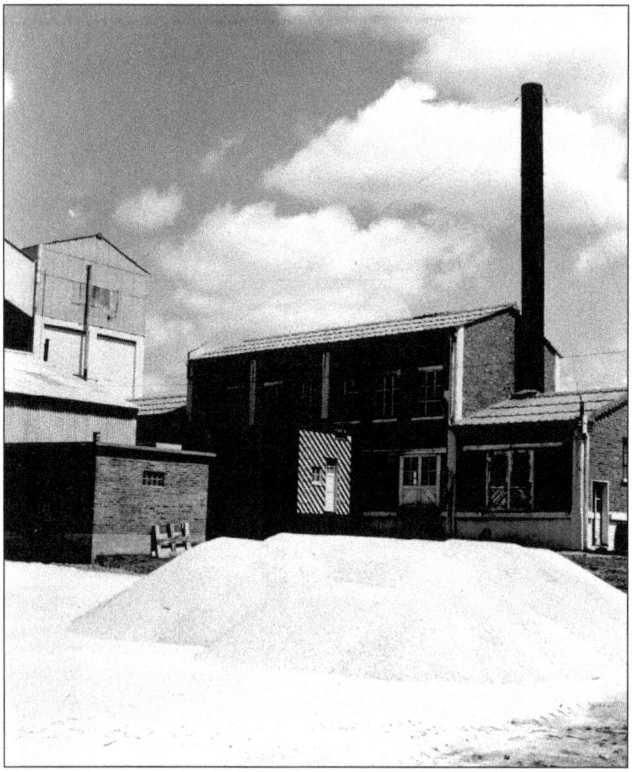

Uneven demand for salt throughout the year proved a continuing challenge. In the early 1950s, excess salt was stored at the inactive Lyons mine, an inconvenient 32 miles away. In 1955, Carey Salt attempted to stockpile its surplus in the open air. According to the March 1956 board of directors report, "The loose salt stored on the pad located east of the truck loading building did not prove satisfactory."

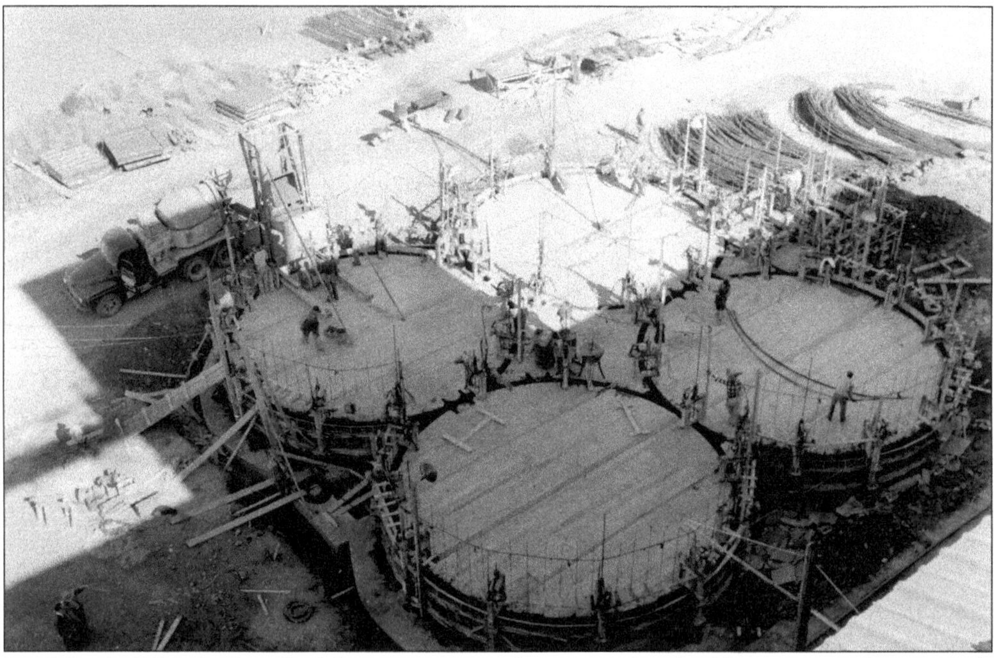

On July 31, 1956, the board of directors approved a proposal to build a concrete elevator with four silos measuring 30 feet by 80 feet high to serve as reserve storage and enable trucks to access salt in the hours when no employees were working in the mill. This image shows the iron reinforcements on the ground, waiting for workers to place them into the concrete forms.

The elevators were built by the Chalmers and Borton Company, a Hutchinson-based firm. Chalmers and Borton is universally known for perfecting the slip form process of building, in which concrete is poured into a form. Once the concrete dries, the forms are raised by a series of screw jacks and yokes, and concrete is poured again. This photograph shows the four-foot form at the top of the partially finished elevator.

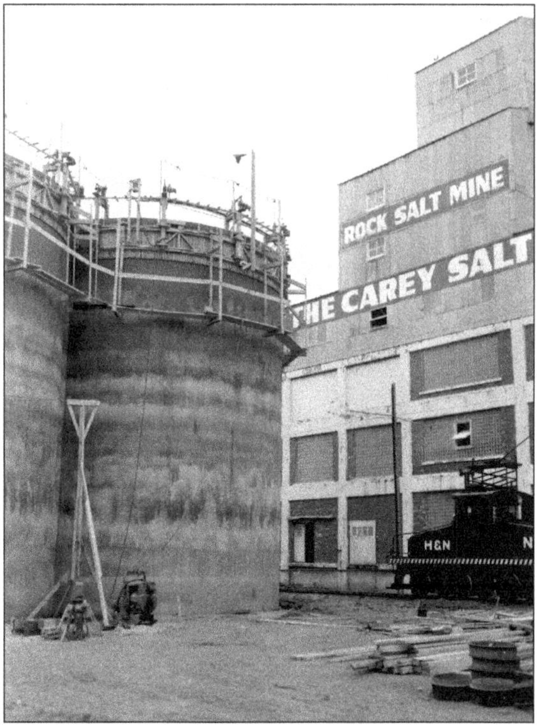

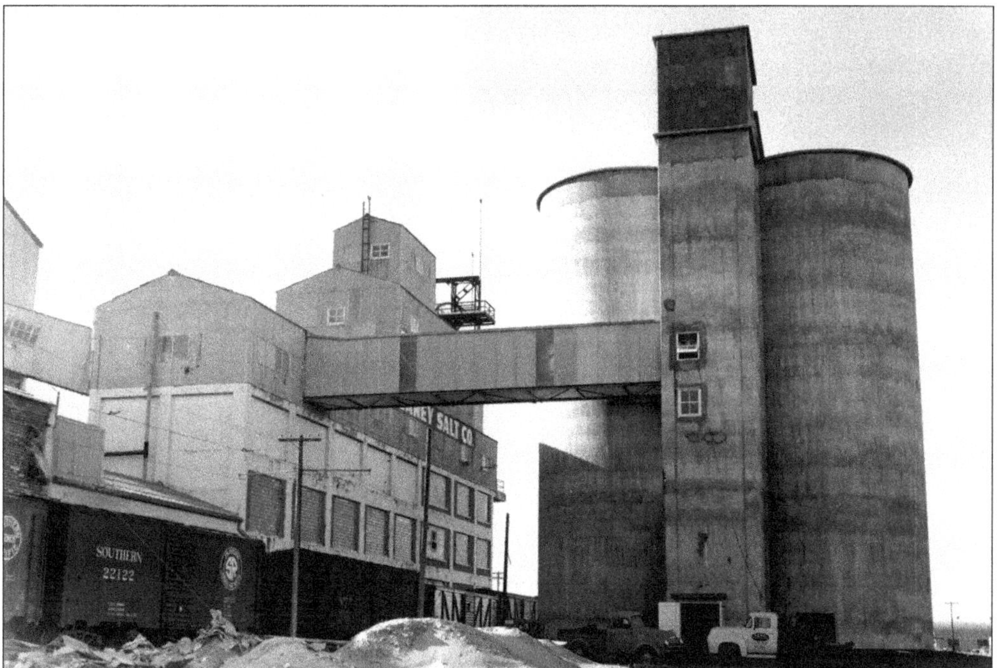

This photograph shows the completed elevator, which was a welcome addition to the mine and liberated the workers from many of the problems of supply and demand. The humidity in the elevators had a tendency to cake the stored salt near the bottom of the elevators, however, and a system of removing the salt from the top had to be devised.

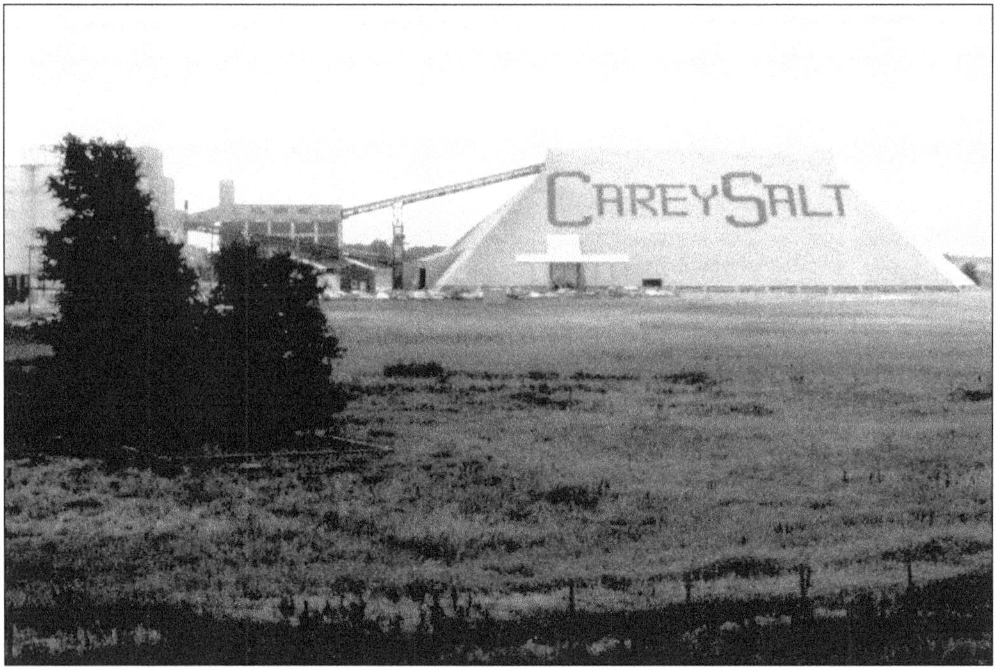

By 1969, the administrators were again frustrated by difficulties in filling winter orders and in having too much salt in the summer. Just after the Interpace Corporation purchased the mine, sales department head Carroll Sargent convinced the new administrators that a storage shed with truck-loading capacity would triple the mine's shipping capabilities and provide continuing employment for the miners. They agreed, and the building was completed in 1970.

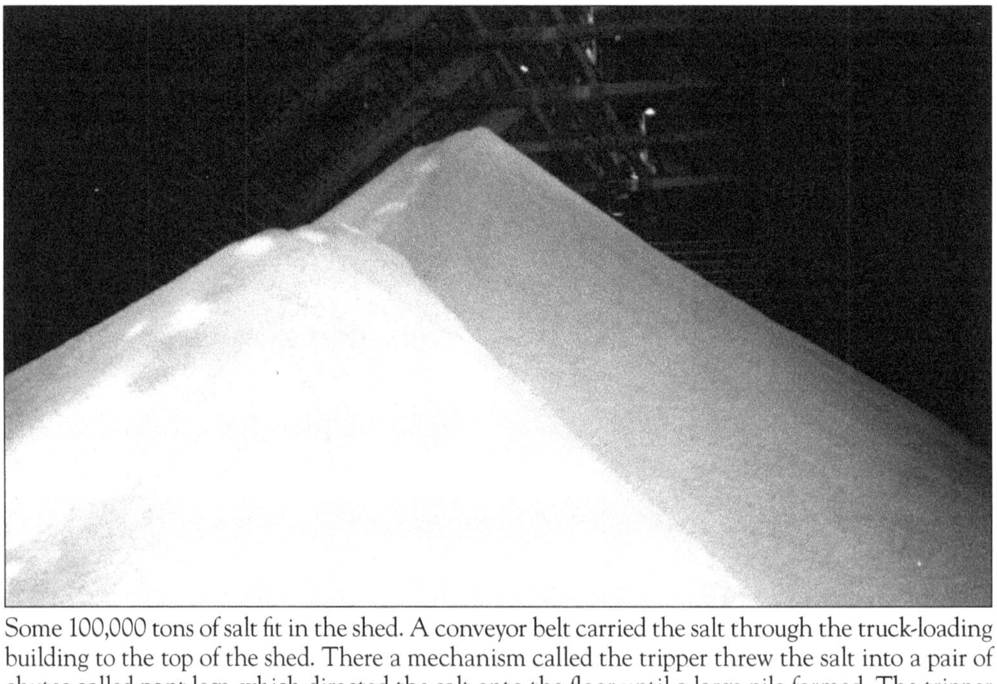

Some 100,000 tons of salt fit in the shed. A conveyor belt carried the salt through the truck-loading building to the top of the shed. There a mechanism called the tripper threw the salt into a pair of chutes called pant legs, which directed the salt onto the floor until a large pile formed. The tripper moved along the length of the shed, allowing piles to develop in different locations.

Seven

SAFETY

Emerson Carey spent two terms in the Kansas state legislature. While he was there, he helped to secure the Kansas State Fair's location in Hutchinson. More importantly to many, he also secured the inclusion of salt workers in Kansas's first workmen's compensation act in 1911. By many accounts, Emerson Carey had respect for the people who worked for him, perhaps because of his own humble beginnings. This attitude was passed down to his four sons, who took over many of the businesses in Carey's empire. Many employees of the Carey Salt Company can still remember one of the Careys coming to visit them at their workstations on their first day of the job. The Carey Salt Company was a family business, and the Careys tried to extend the feeling to their employees. Each Christmas through the 1950s, Howard J. "Jake" Carey hosted a party at his home for the children of the employees.

Fostering this attitude may have been successful because many of them *were* family. Both the evaporation plant and the mine were known for its comparatively low turnover rate. Between the years 1950 and 1955, the turnover rate was 2 percent, and average tenure was 20 years. Moreover, fully 20 percent of new employees had relatives who were also working for Carey Salt.

The safety of the employees was a priority at all the facilities. In 1932, the Carey Salt Mine and evaporation plant had the lowest accident rate in the country. In the future, the mine would win many awards for its safety practices. Overall, mine safety training was not as structured as it is now. Increasing regulations by MSHA have played a large role in training during the last 30 years.

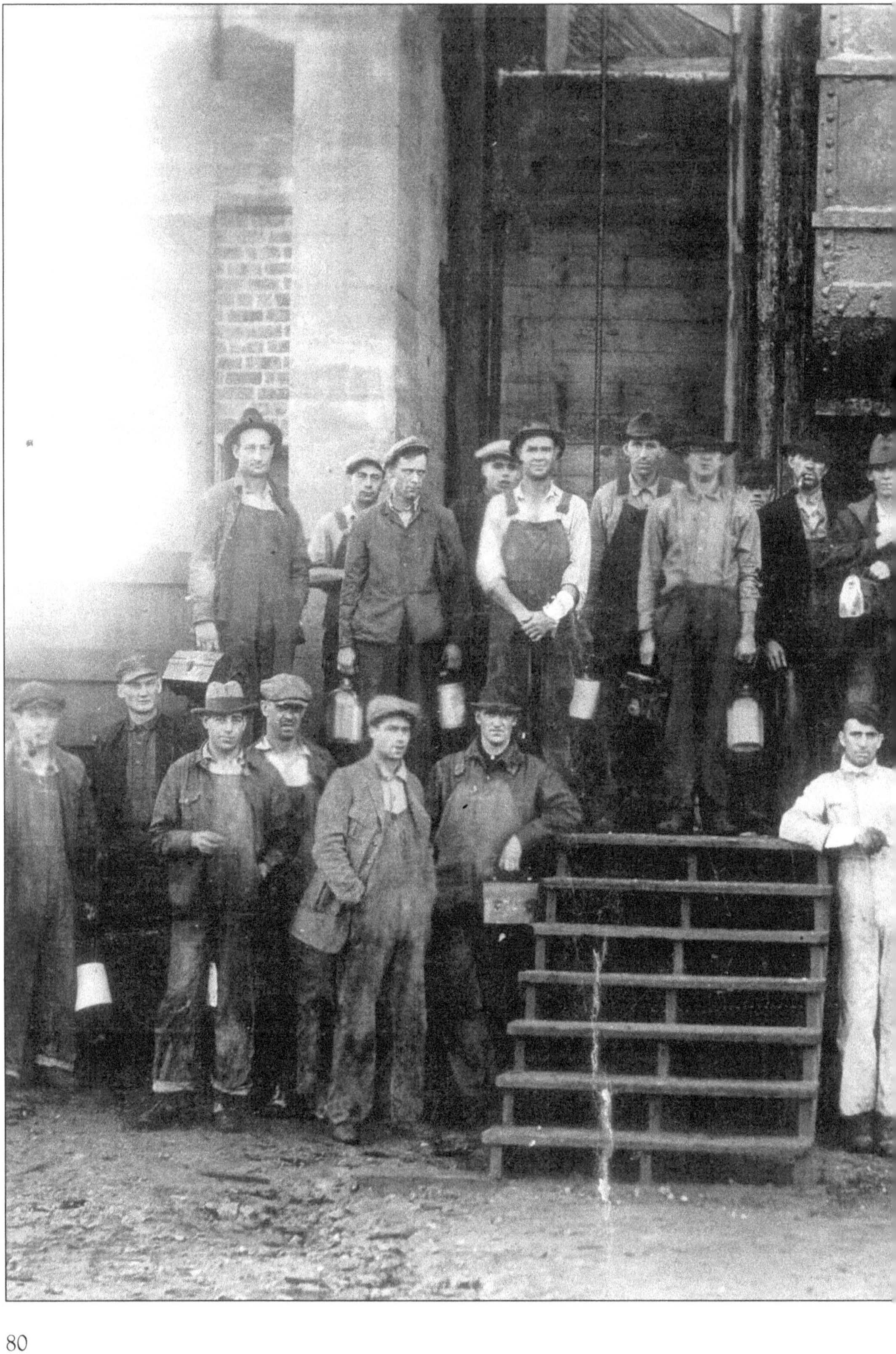

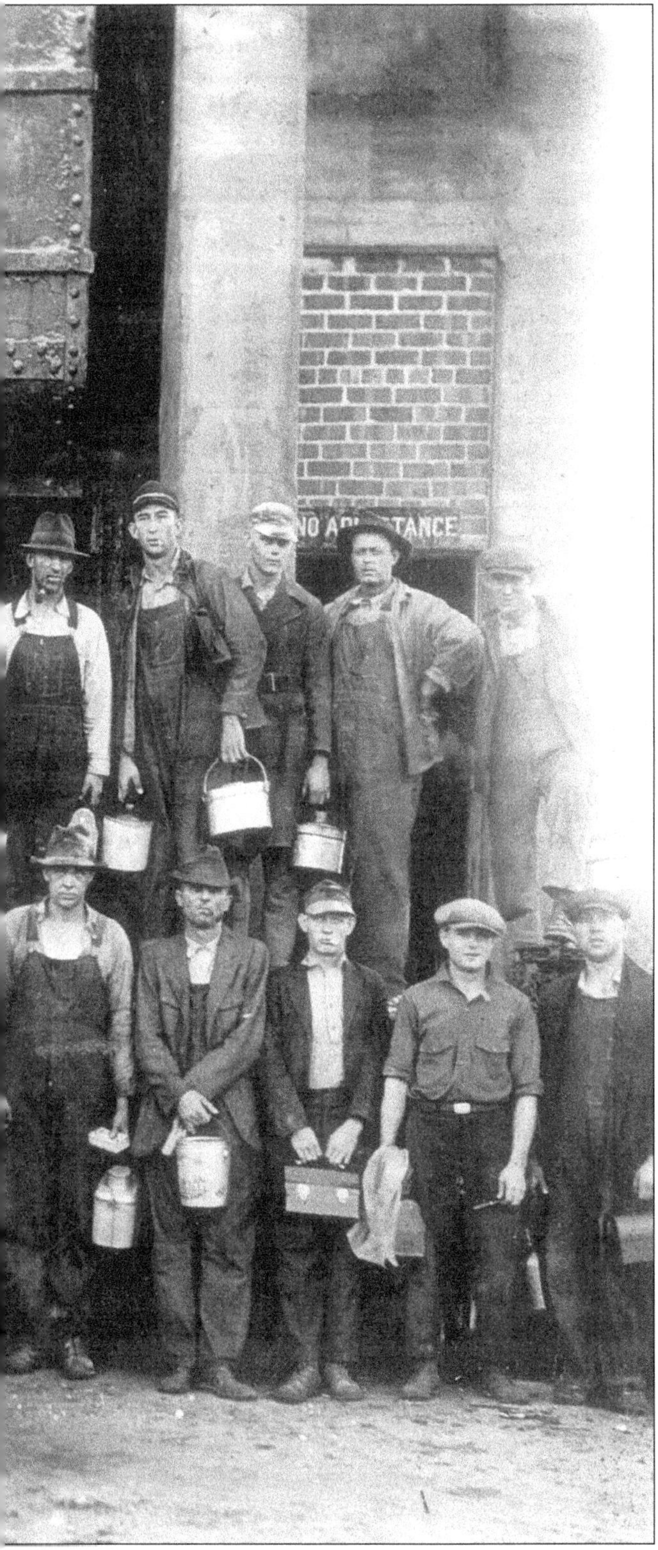

Throughout the years, many Carey Salt Company plant and mine employees were married to one another. Sons followed fathers into the mine. Brothers and in-laws worked alongside one another. In this mid-1920s photograph, Calvin Tracy appears in the back row, second from the right. His son Wallace stands in the back row, fourth from the left. Albert, Calvin's other son, appears fifth from the left in the front row, next to brother-in-law Everett Roberts. All these men eventually worked themselves into manager or superintendent positions. But this may not have been their greatest contribution to the company—two more generations of the family went to work for the company in the following decades. (Courtesy George E. Roberts.)

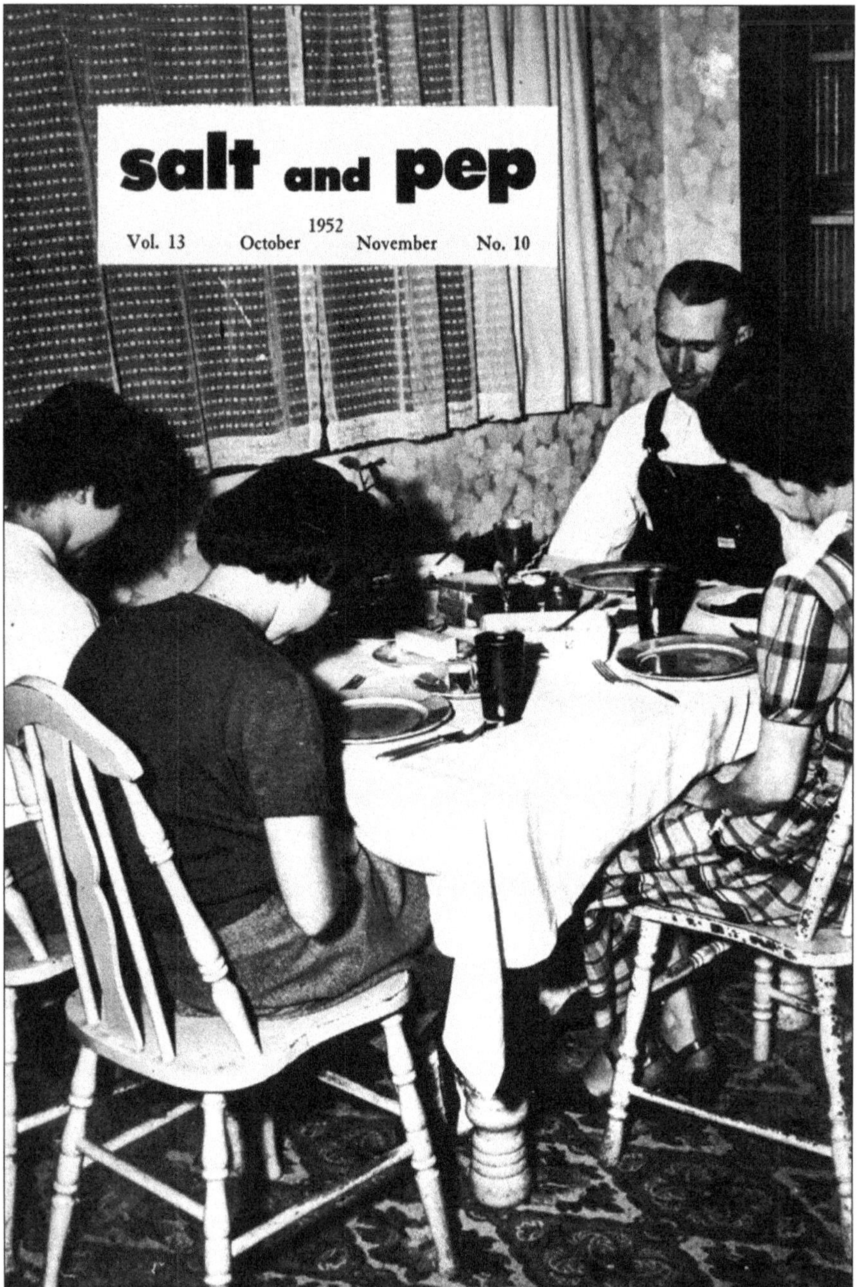

The periodical *Salt and Pep* ran for several years as a magazine where employees could catch up not only on work-related issues but also personal topics such as marriages, vacations, and new employees. In 1952, miner Raymond Law and his family were featured in the Thanksgiving issue, pictured here. An example of the family feeling between employees occurred in 1953, when miners voluntarily rearranged their vacation schedules to delay seasonal layoffs for 12 of their coworkers. This sense of camaraderie was particularly strong before the Carey family sold to the Interpace Corporation in 1969. Even today, former employees of the Carey Salt Company meet monthly to eat, drink, and gossip.

Most of the main facilities of the Carey Salt Company had bowling and basketball teams. Members of the Carey family were devoted to the game of golf and sponsored annual sodbuster tournaments for their employees. These extracurricular activities solidified the feeling of belonging for Carey employees. This 1952 photograph shows the miners' bowling team. Pictured are, from left to right, Jake Herdt, Wes Swartzell, H. Bush, Earl Bush, and Edward Haines.

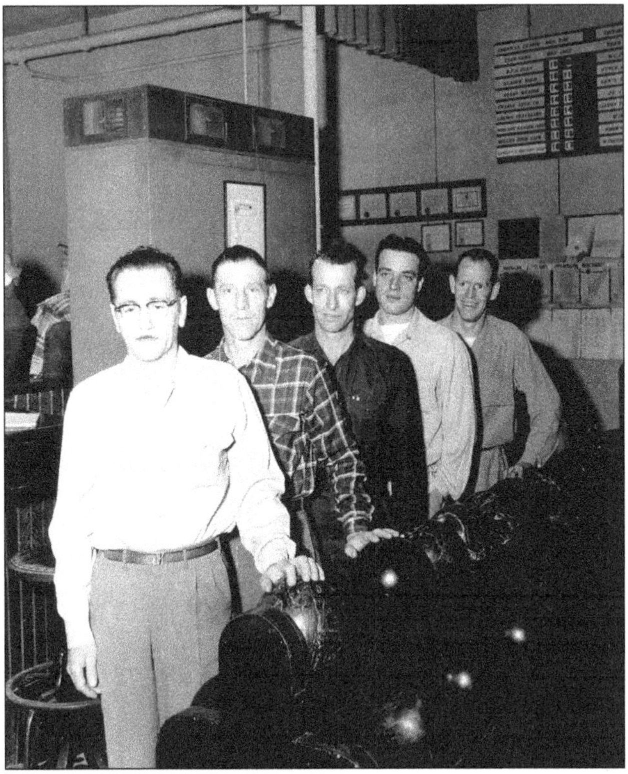

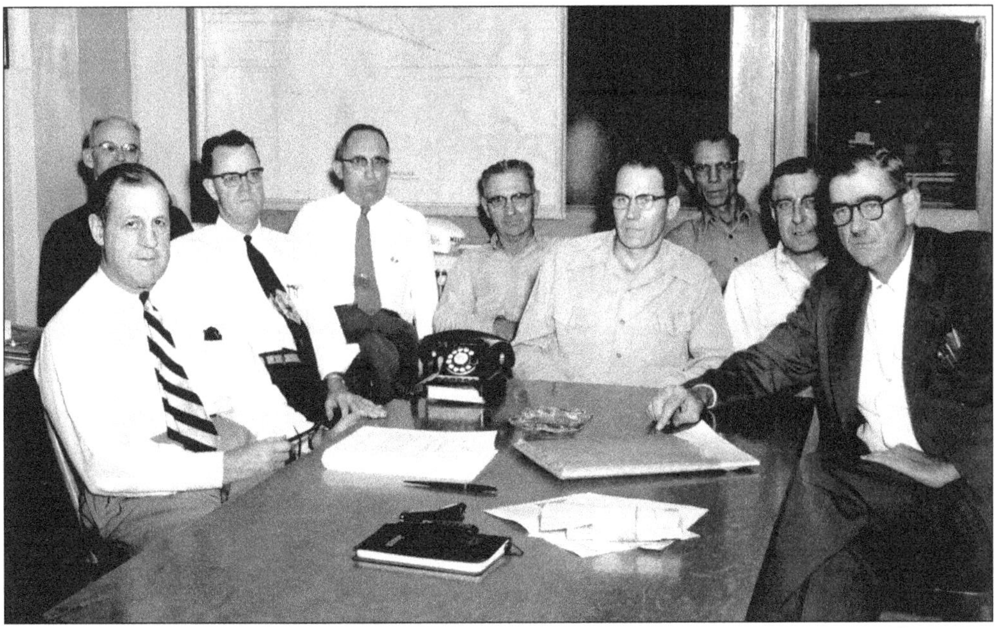

In 1941, the Hutchinson miners succeeded in unionizing under the United Mine Workers after an earlier attempt failed in the 1930s. By 1956, union representatives were asking for a 40-hour workweek and overtime pay, a wage increase of 15¢ per hour, and hospitalization insurance for senior employees. This photograph shows union representatives from the Hutchinson mine and the evaporation plant signing a contract renewal for 1954–1956.

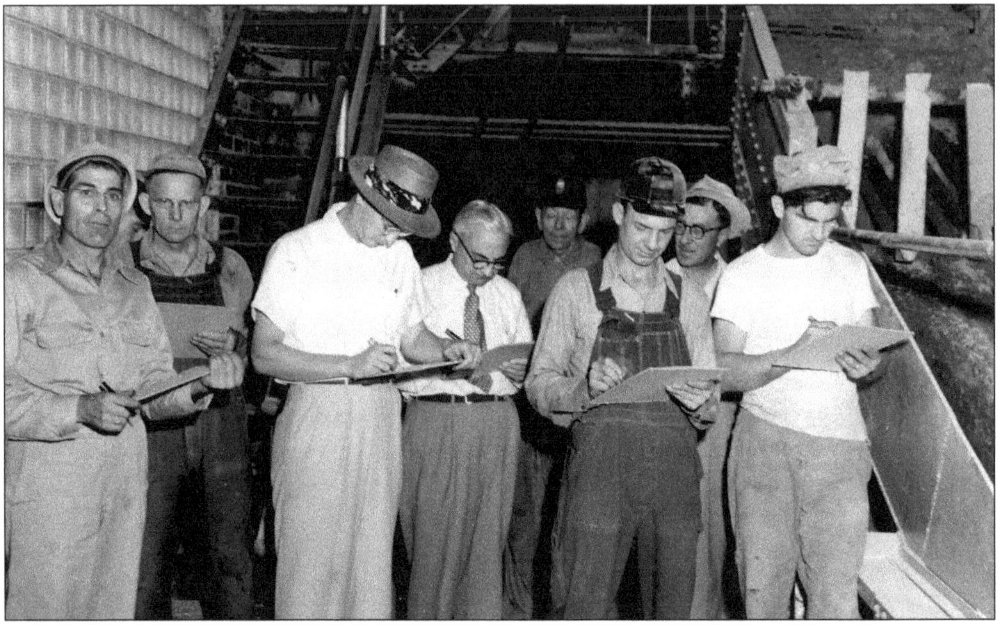

Regular safety meetings with representatives from each plant took place at least from the mid-1930s. Guest inspectors from the corporate office, the plant, and the mine went on inspections to make observations and target potential safety hazards. This 1951 photograph was taken in the mill building at the mine.

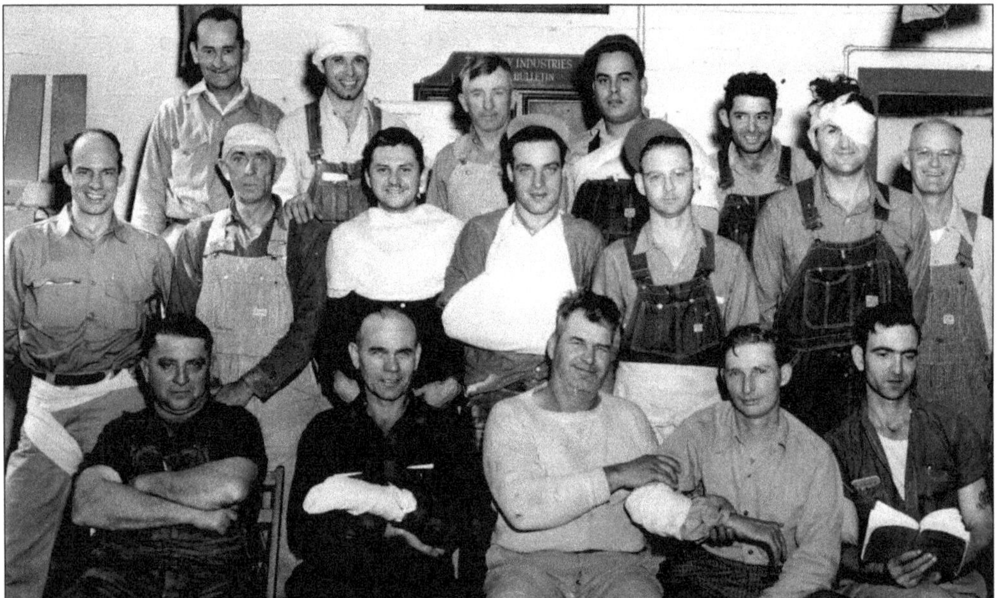

All employees were required to take a first aid class given by instructors who had been trained by the U.S. Bureau of Mines. Seen here are, from left to right, (first row) Rudy Philbrick, Abe Janzen, Clinton Flanders, Willie Thrift, and unidentified; (second row) Everett Cline, Charles Hungerford, Joe Schumacher, Earl Bush, Charles Chalmers, and Vernon Horton; (third row) Clyde Hackworth, John Thiessen, Claude Ryan, Bacho Rodrigues, LeRoy Badders, and Bennie J. Pallister.

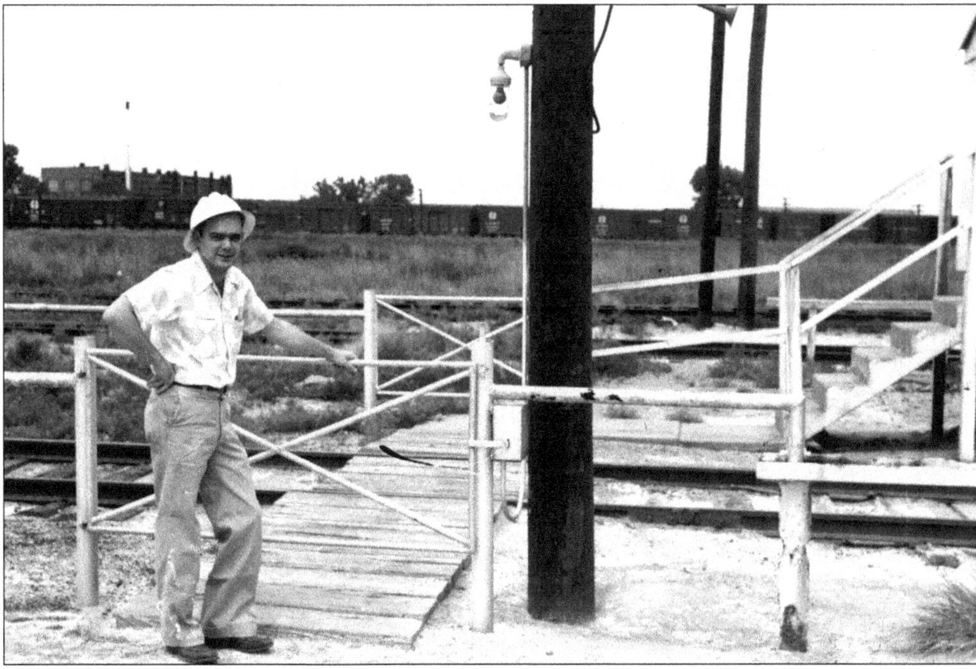

The Carey Salt Company's safety campaign was not restricted to belowground or to the mill. Superintendent Ron Stone poses here at the new safety gates blocking the pedestrian sidewalk across the rail tracks in this 1954 photograph.

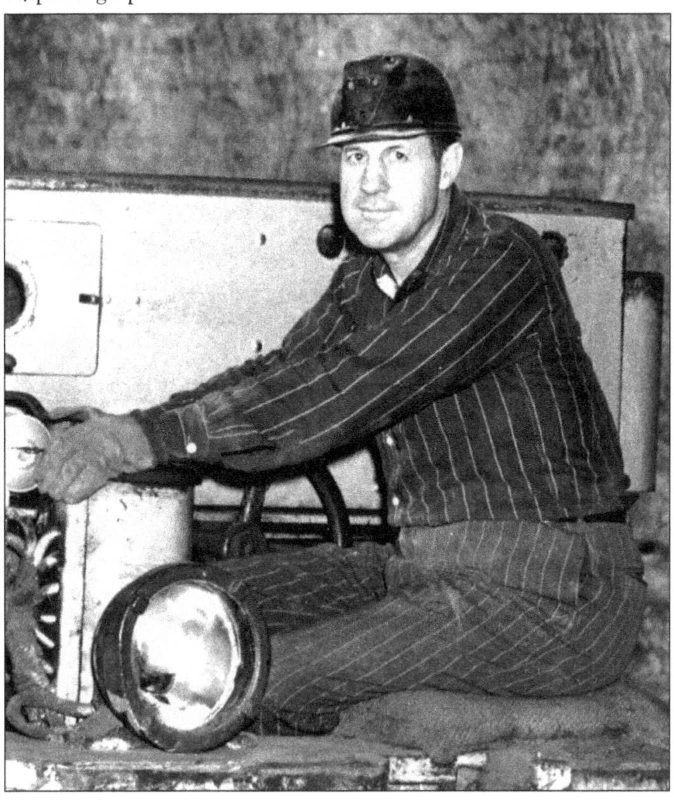

Steven B. Horrell, who began working at the mine at about the time it opened, made his way up the ladder to become vice president of operations. He was responsible for much of the safety awareness in the mine and was proactive in many issues, especially regarding safety gear. A pioneer of safety in the industry, Horrell was eventually made a director of the National Safety Council.

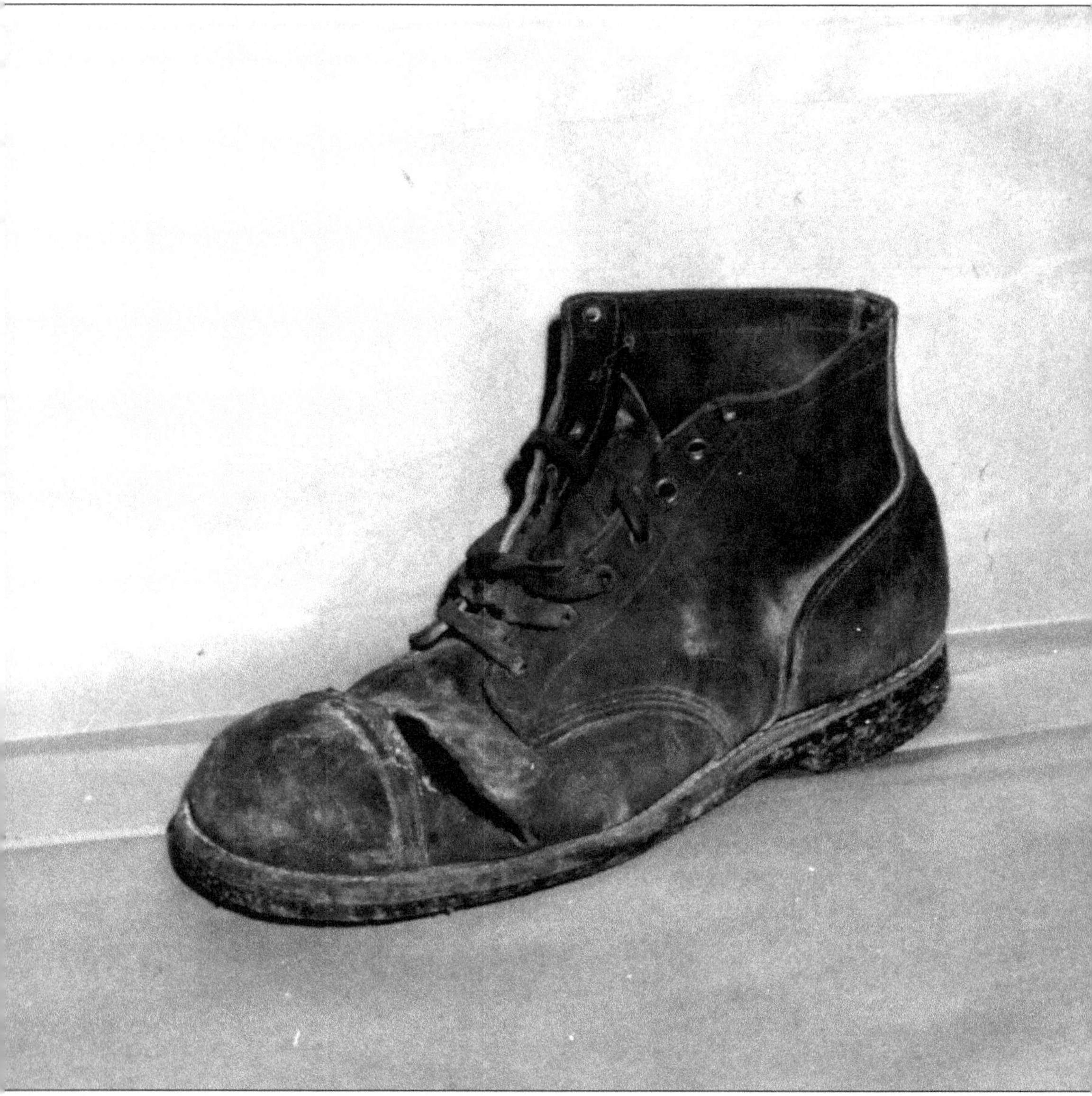

Miners could buy safety gear from the Carey Salt Company, but for many years, it was not required. In 1935, Steven B. Horrell wrote to the International Shoe Company, complaining that the leather of the shoe was unable to withstand the abrasive salt and would rip, loosening the steel toe box. The company responded by developing a shoe that would withstand the conditions in the mine, available to the Carey employees for $3 per pair. This picture appeared in the July 1952 issue of *Salt and Pep*, highlighting an accident in which shoes saved a miner's foot from being crushed. The article went on to say that the miner "has been in the hospital where two of his toes had to be amputated, but is progressing nicely. . . . It's a heck of a lot better to have three toes than to be walking on a stump for the rest of your life."

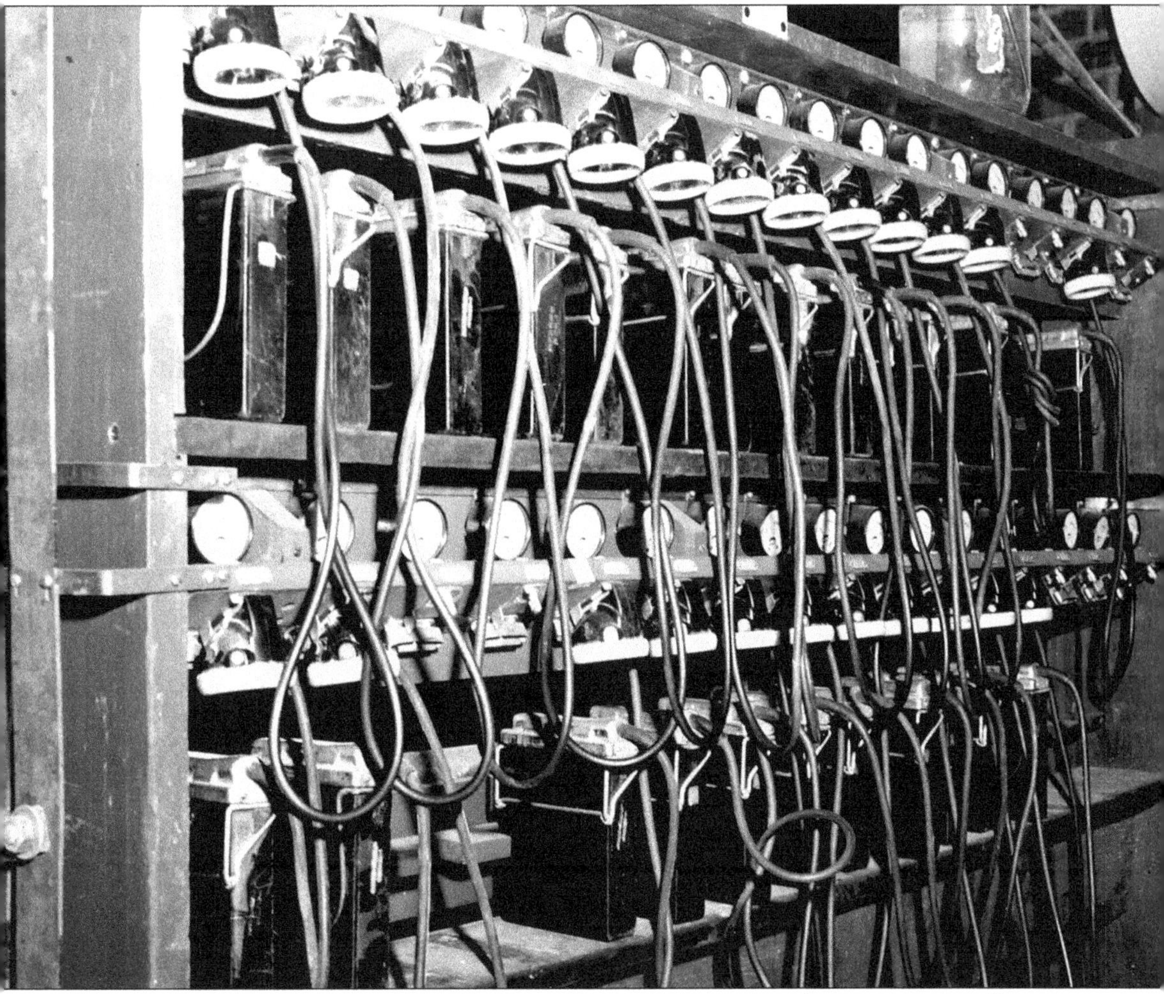

At least until the mid-1940s, the miners wore carbide lamps on their hats. Bennie J. Pallister's memoir indicates that the lamps were hazardous and could interfere with working relationships. "At times when we needed to work close together, perhaps using a jack to put the locomotive or a car back on the track after a derailment, suddenly there would be a smell of burned hair, generally followed by a little blue smoke, caused by some unkind words." He also had this to say: "I quickly learned one lesson, never to carry matches [for the lamps] in my pants pockets. . . . One day while riding on the bumper of two cars, ... I rubbed against the car end and set a bunch of matches on fire, resulting in somewhat of an Apache war dance and a burn about the size of a half dollar. Matches were then carried in a box in a shirt pocket." By 1951, the carbide lamps had been exchanged for battery-operated lights, shown here in their chargers.

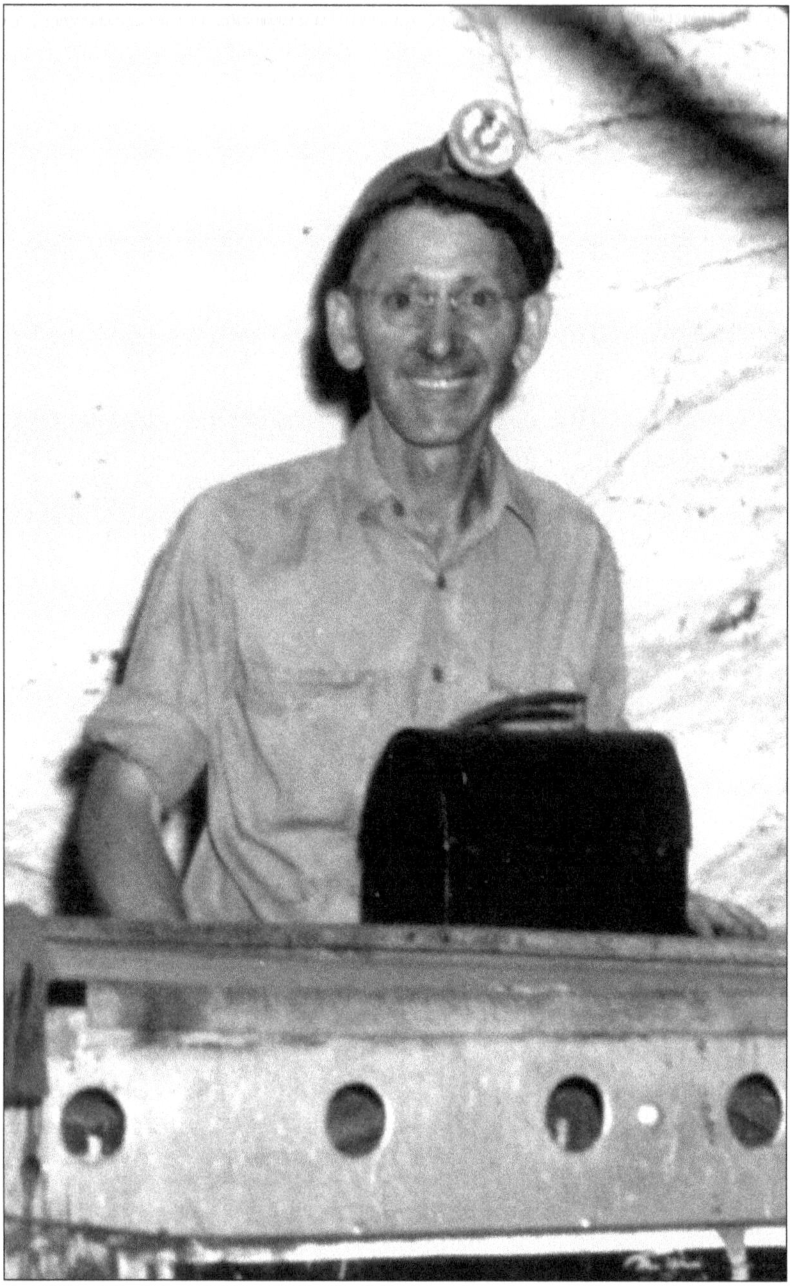

An article in the November 1951 issue of *Salt and Pep* laments that while the Hutchinson and Winnfield mines worked a total of 135,281 man-hours without a disabling accident during 1950, smaller accidents were on the rise by 29 percent. "Each upsurge in the number and seriousness of accidents is cause for alarm on the part of workers and management. It calls for redoubled effort." That very month, powderman Bill Tilton lost his life in an explosion underground. He was the first Carey Salt Company employee killed in a mine-related accident. Then in June 1953, Fred Swepston was caught between two colliding locomotives and killed. Swepston is pictured here in this 1952 photograph. Since the mine opened, there have been only two mining-related deaths underground at the Carey Salt Mine.

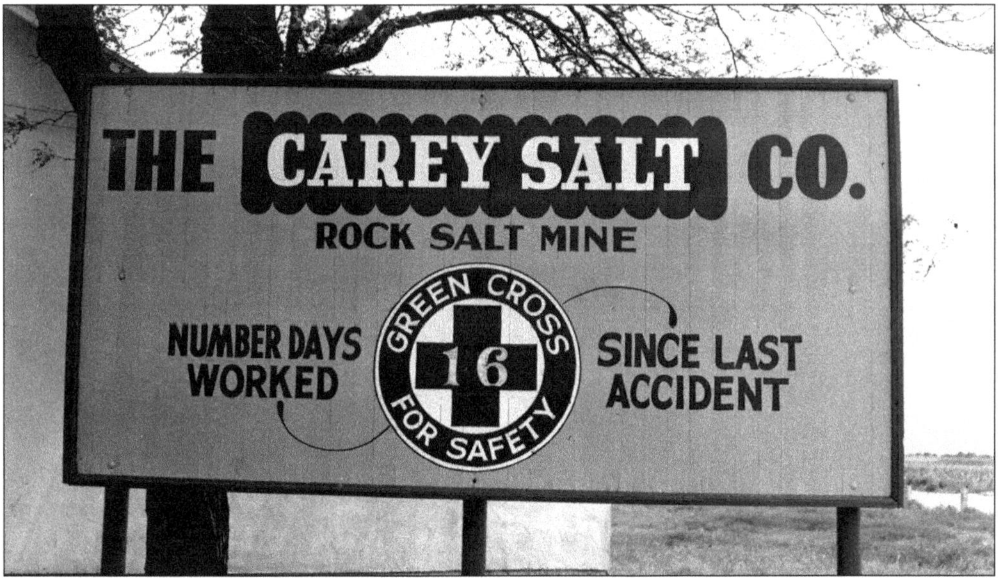

The Carey Salt Company offered several incentives for safety on the job. All who served one year without a lost-time accident received an extra day's pay. The second year, the miners would receive two days' pay. In 1951, miners received four days' extra pay for their four years' record.

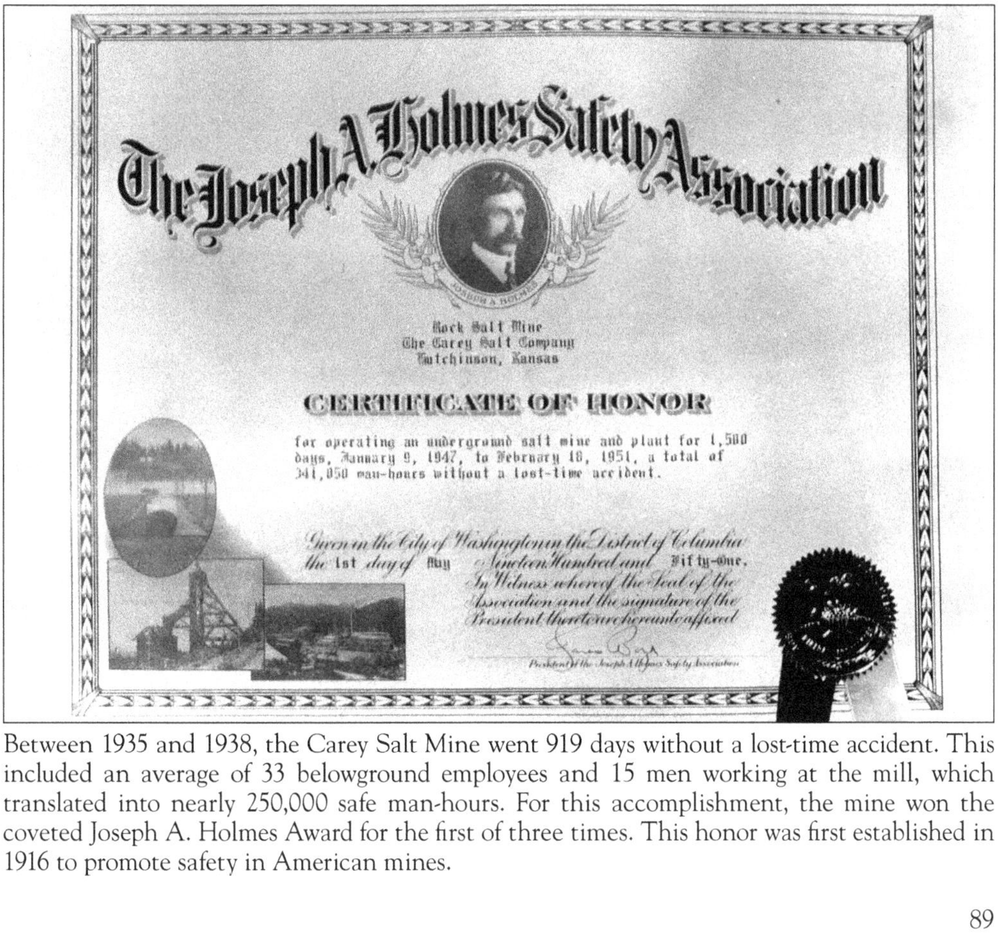

Between 1935 and 1938, the Carey Salt Mine went 919 days without a lost-time accident. This included an average of 33 belowground employees and 15 men working at the mill, which translated into nearly 250,000 safe man-hours. For this accomplishment, the mine won the coveted Joseph A. Holmes Award for the first of three times. This honor was first established in 1916 to promote safety in American mines.

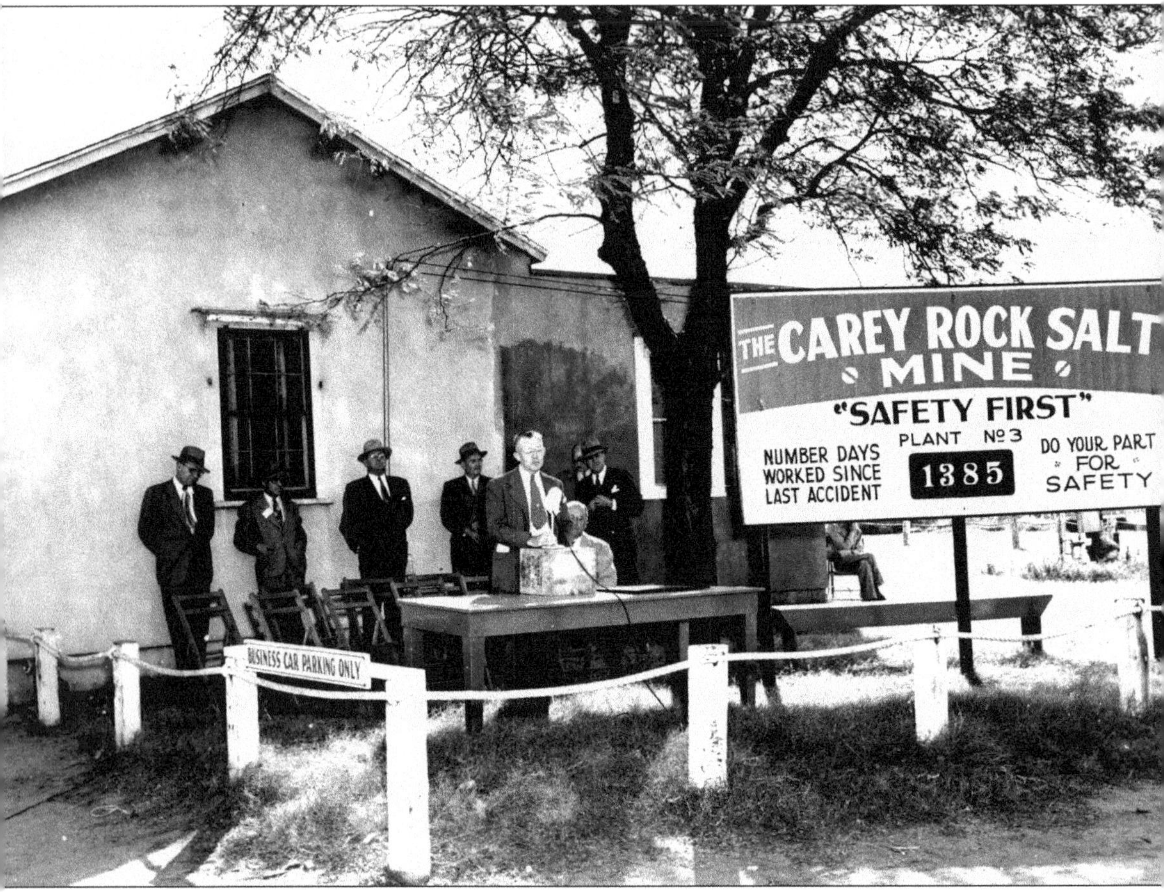

In October 1950, R. D. Bradford from the U. S. Bureau of Mines honored the Carey Salt Mine for 1,385 safe days. Then in February 1951, the Carey Salt Company once again received the Joseph A. Holmes Award. In the years between 1982 and 1984, the Carey Salt Mine received awards from the Salt Institute for perfect safety, and in 1984, the mine achieved 100,000 man-hours without a lost-time accident.

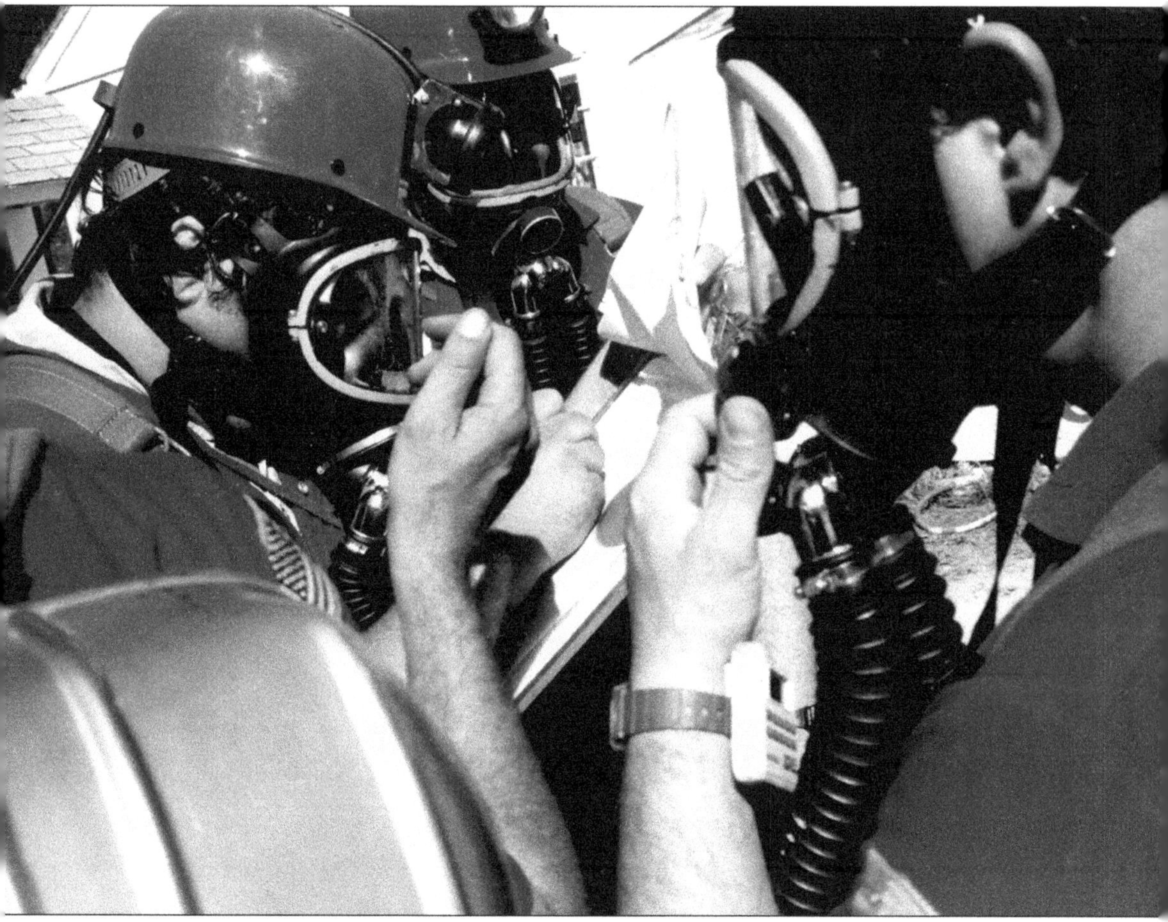

MSHA requires that each mine have a rescue team in the event of a disaster at its own or a neighboring mine. To encourage their training, mine rescue teams vie for awards at competitions. The year 1983 marked the first time the Carey Salt Company sent a team to a national competition. In this 1987 photograph, mine rescue team members, from left to right, Myron Marcotte, Bob Smith, and Mike Harra prepare a rescue strategy before entering the mine in a mock disaster contest at Rolla, Missouri.

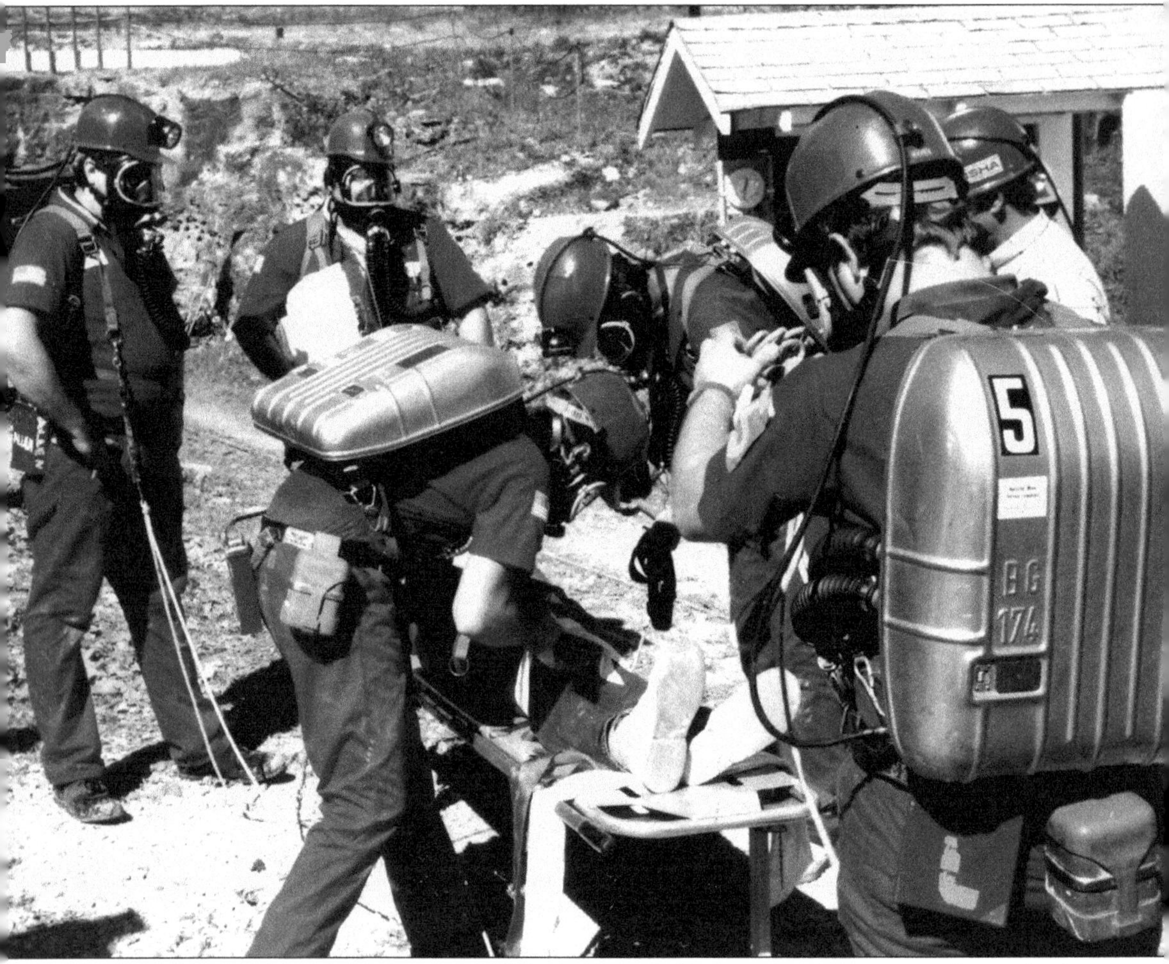

Training, thanks in part to new MSHA regulations, has increased exponentially in the last three decades. According to Myron Marcotte, longtime employee at Carey Salt and, later, the Hutchinson Salt Company, "When I started [in 1976], they showed me a self rescuer, told me what it was for, and put me with an older miner for a couple of days. That was my training. Now we have to give 40 hours' documented training to each new hire, then we have to put them with a miner for the job they will be doing. Then everyone has to have eight hours of annual refresher training." This photograph shows Carey's mine rescue team working on an "injured" miner at the 1987 competition in Rolla, Missouri. Pictured are, from left to right, Alan Roberts, Myron Marcotte, Bob Smith, Guy Hagan, and Mike Harra.

Eight

VISITING THE MINE

Nobody really remembers a time before visitors were allowed underground. As early as 1927, a man-cage hung directly beneath the skips, more likely for the prospect of bringing visitors underground than for handling the miners, who rode in the buckets in other mines. Rumors of the fancy dinner parties that the Careys hosted underground may or may not be true, but the large groups that tromped through the mine by the hundreds during the day are well documented. Groups of all kinds could call ahead and take a tour of the mining operations.

This practice was discontinued during World War II, probably due in part to the labor shortage and the extreme demand of salt for the war effort. The tours revived again soon after the troops returned, however, much to the dismay of the foremen underground. The administration evidently felt that experiencing the mine face firsthand would make an impression on visitors that would pay off when they went to the store, compensating for any inconvenience that the miners might encounter. This may have been true, for many people still remember going underground as schoolchildren in the 1940s, 1950s, and 1960s.

In 1961, an article in *Salt and Pep* stated that the Carey Salt Mine would no longer take visitors because the extreme popularity was making the tours too difficult to handle. Doubtless increasing anxiety about public safety and liability issues played an additional role in discontinuing the practice. Not long afterward, an article in the *Hutchinson News* proposed sinking another shaft into the mine to accommodate tourists, so popular was the experience. Finally in 2007, the Reno County Historical Society turned that idea into reality when it opened the Kansas Underground Salt Museum, allowing visitors to see the caverns once again.

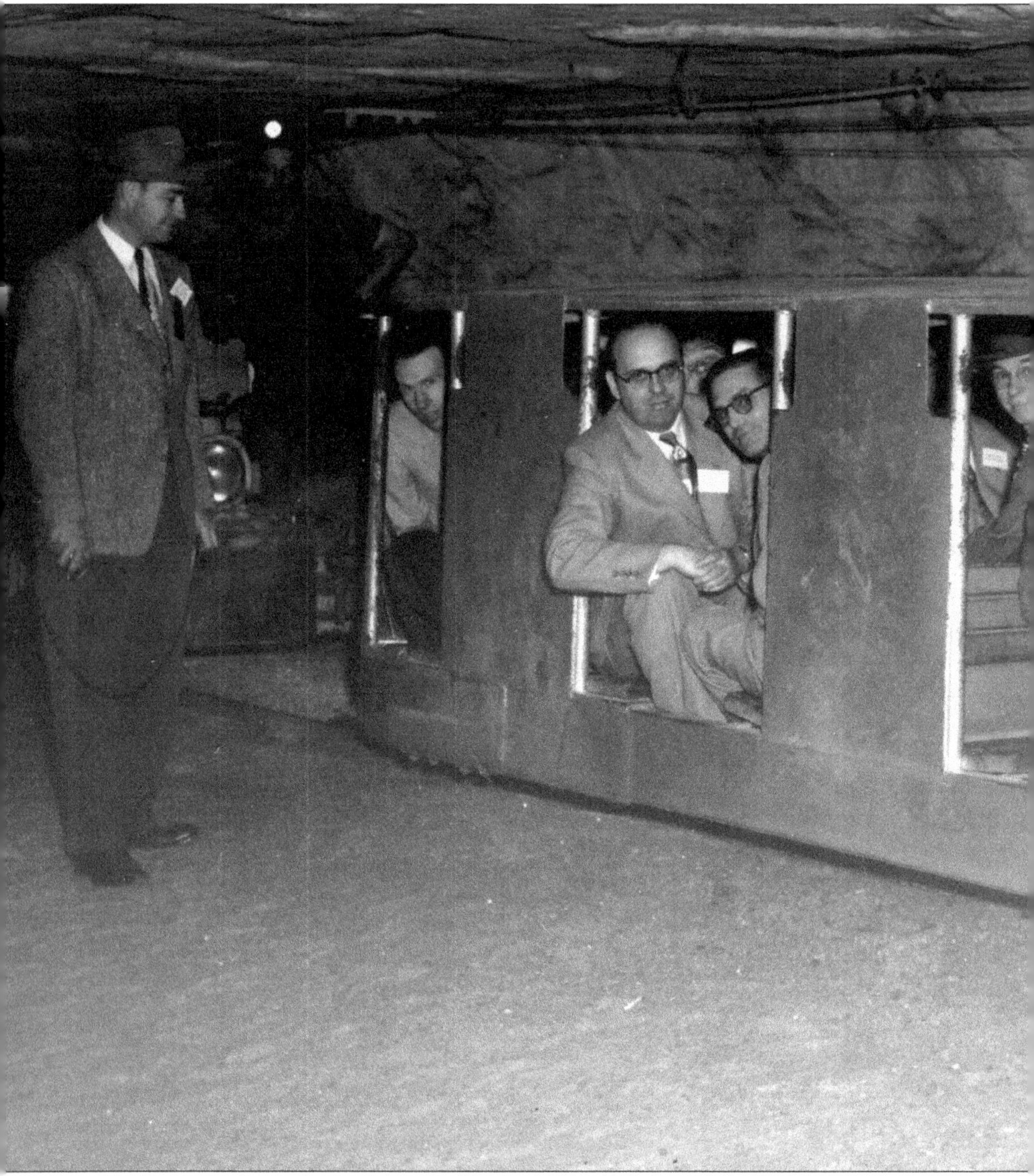

"Climb aboard the specially designed man car which was constructed to ensure the safety of visitors . . . and mine personnel. . . . We pull out on the main track and down the mile long haulageway. The eerie screams of the siren echo through the dimly lit corridors. We pass the powder room (where, to the dismay of ladies in the party, only dynamite is stored) and numerous tunnels that have been blocked to give proper direction to the fresh air that is circulated through

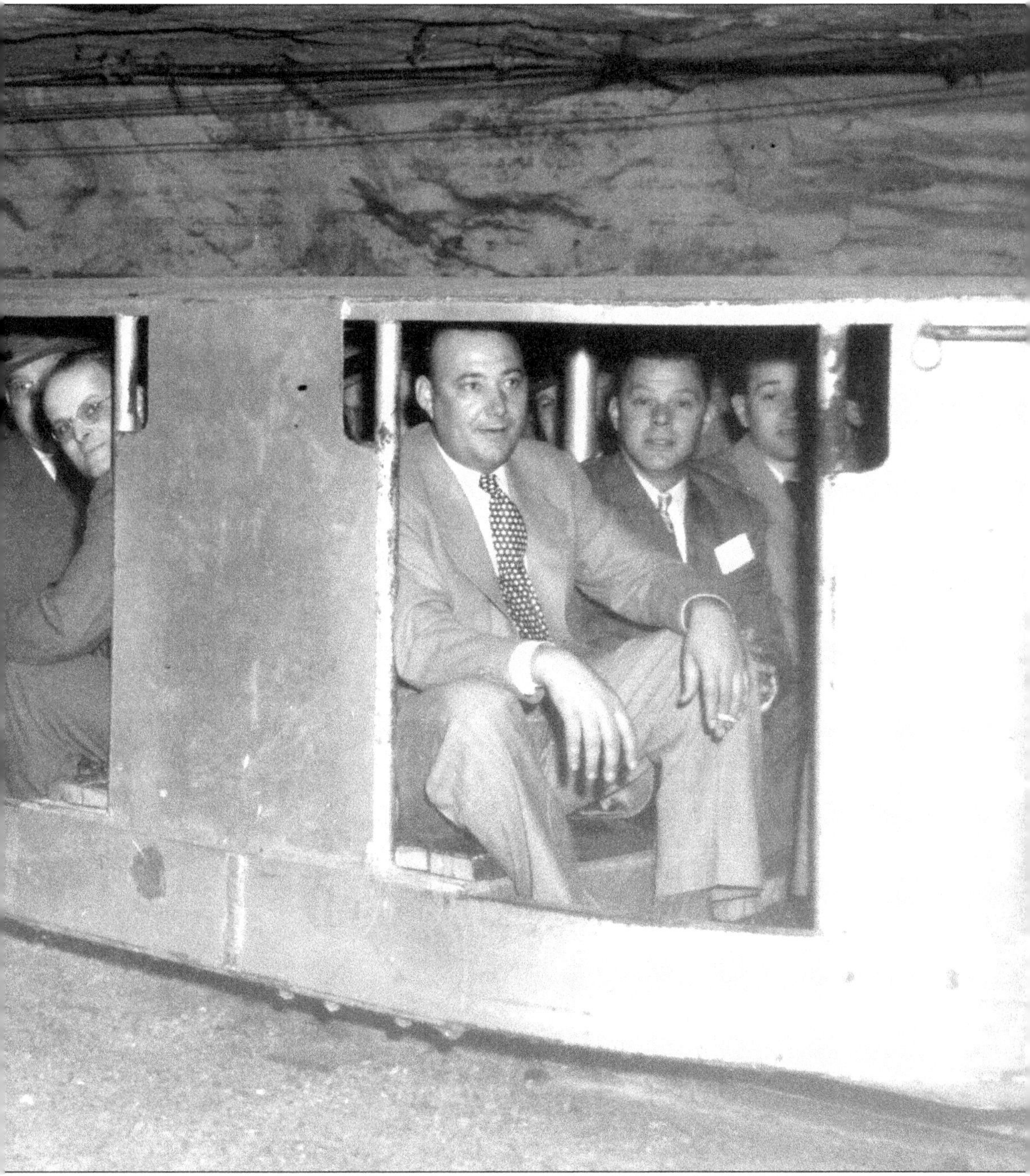

the mine. A sharp curve and our train comes to a stop. We de-train at a loading station near the working face." Members of a special party underground read that passage from the booklet *Lunch in a Salt Mine*. In 1950, these employees of J. S. Dillons, a chain of grocery stores, traveled underground to see the mining operations.

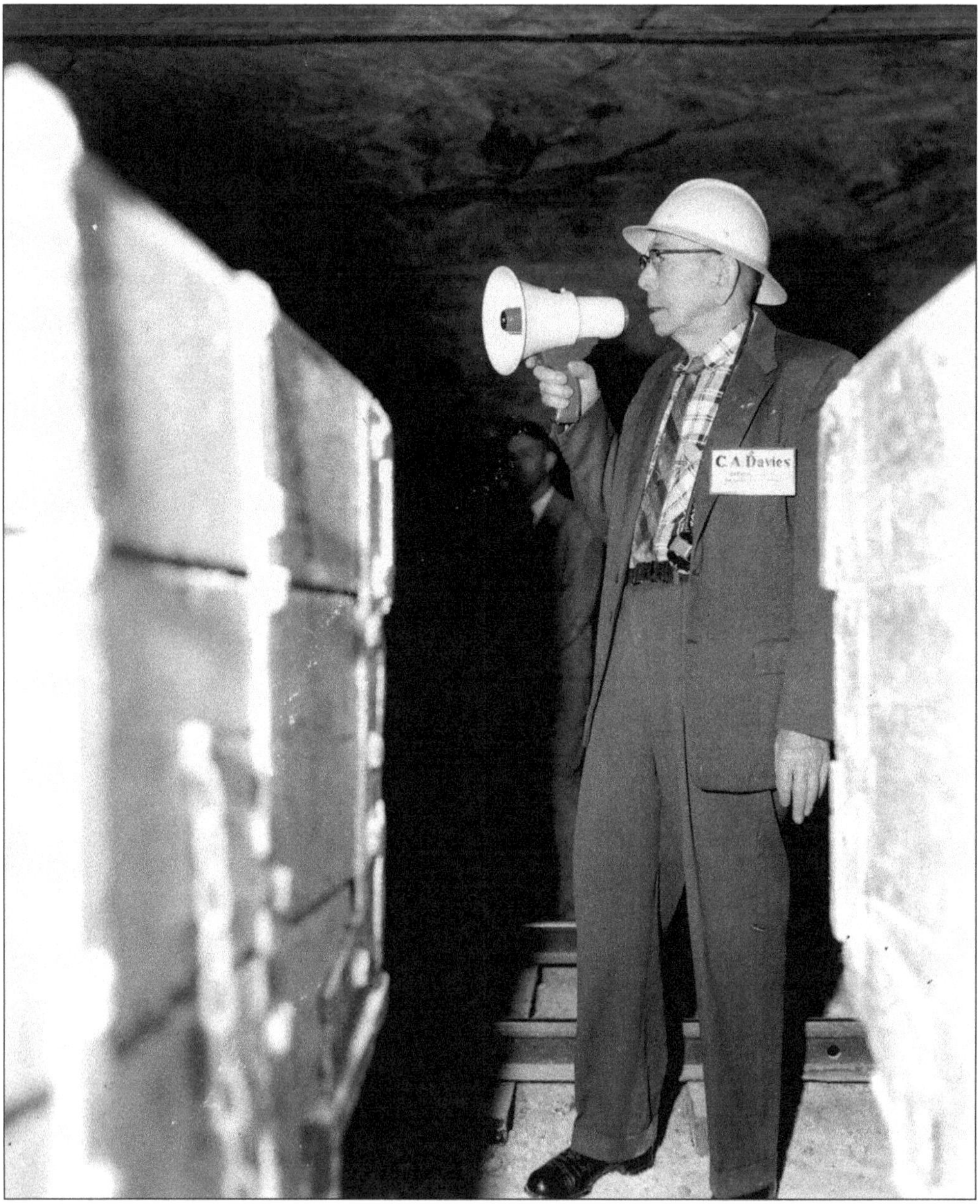

Bud Davies was an electrician at the mine for many years. After he retired, he returned to give tours to visitors. A 1958 requisition shows that this portable loudspeaker was purchased for $92 because the noise level was so high in the mine that nobody could hear the guides.

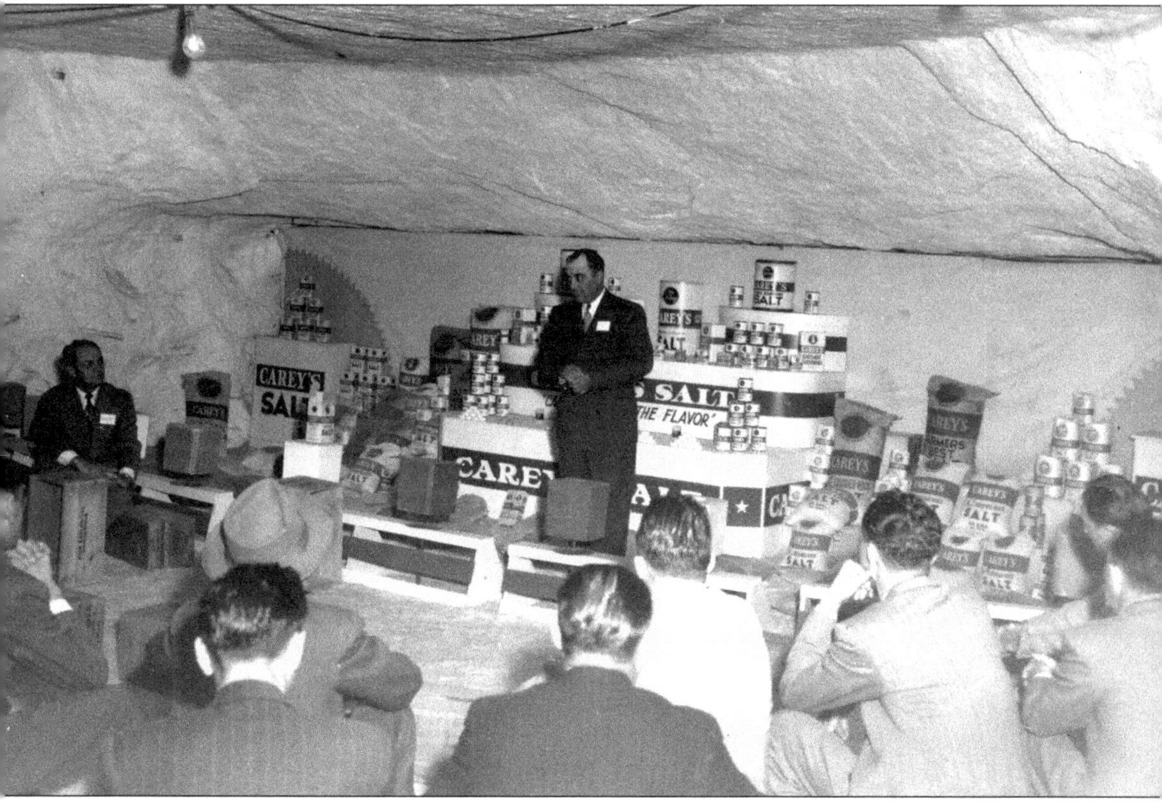

The underground experience often ended in a room with a Carey Salt Company trade show display. Giving people the opportunity to see the mining operations was an important marketing tool. A 1959 *Salt and Pep* issue reminds the miners, "The time and effort spent in arranging and conducting tours is well worthwhile and plays an important role both in advertising and selling of all types of Carey Salt products. The tour program permits you, who actually process salt to become valued assistants to the Sales Division. You are on the right spot at the crucial moment, and at this point can influence a visitor to become a buyer of the product."

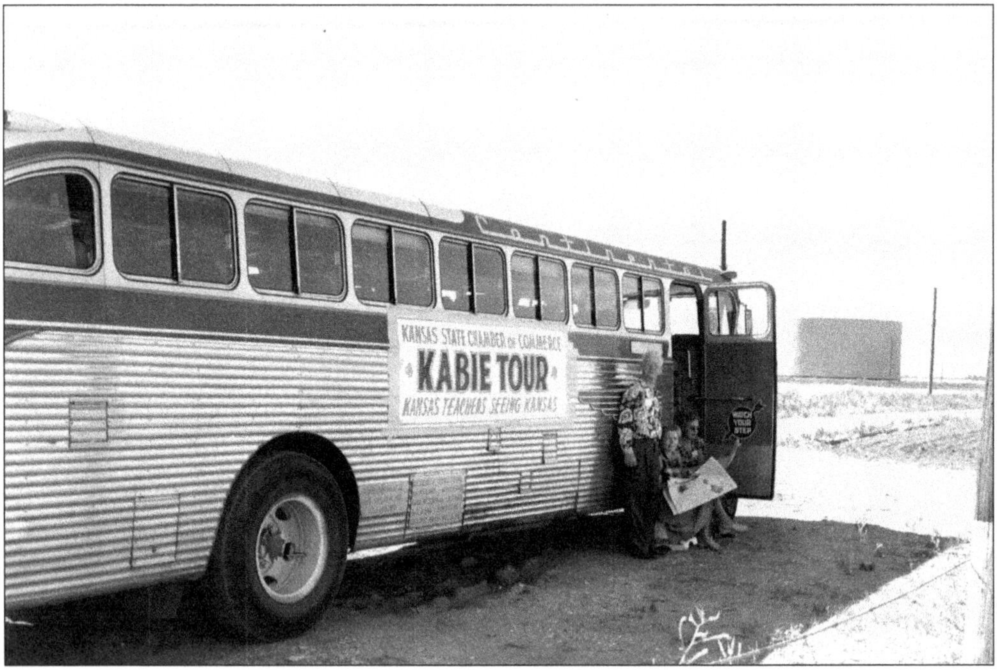

The Kansas State Chamber of Commerce sponsored an annual tour of points of interest throughout the state during the 1950s for Kansas teachers. For many years, the salt mine was a highlight of this event, known as the KABIE tour.

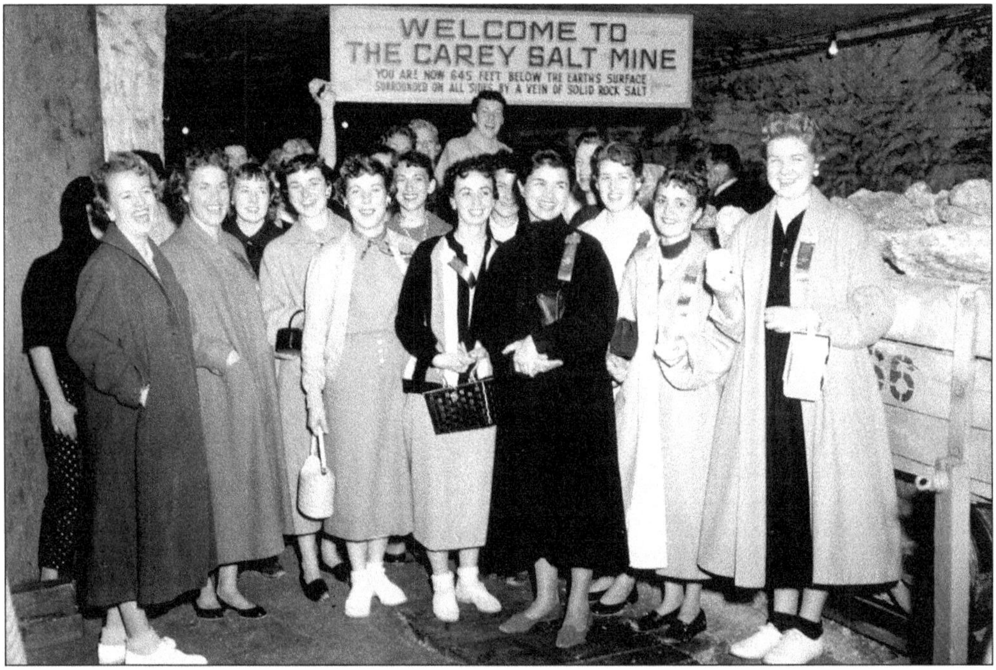

Teachers, like this group, from all over the state participated in the KABIE tours. In 1958, the Carey Salt Company renovated the ladies' restroom. A note on the requisition form stated that "with the heavy burden of visitors we are now handling, we feel we need this repair and recondition." The tours influenced the teachers to bring their students on field trips.

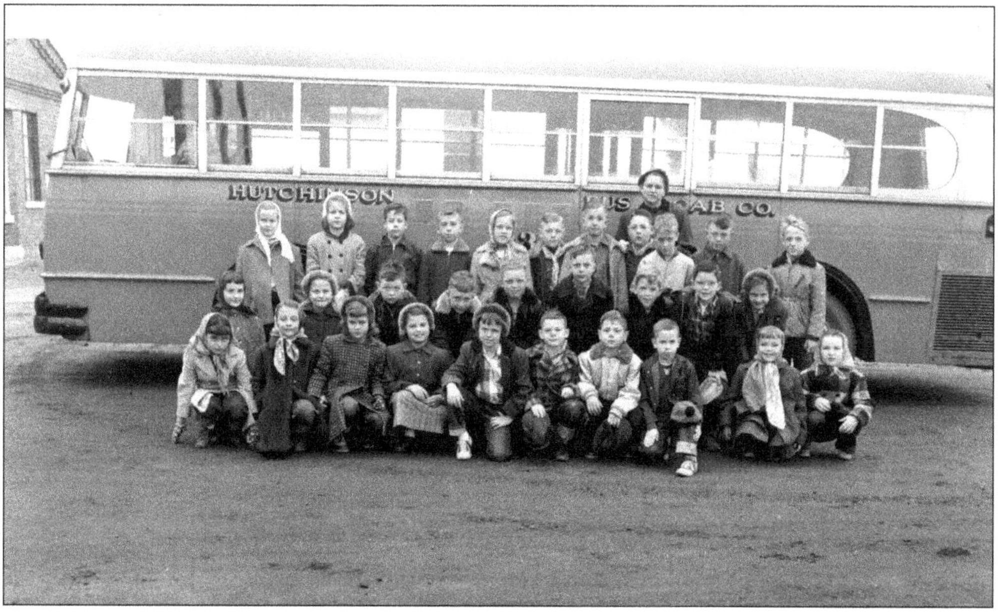

This group of third graders from Hutchinson's Morgan School was typical of many groups that toured the mine. For many students, traveling underground was the biggest field trip of the year, and many adults remember taking the trip as children.

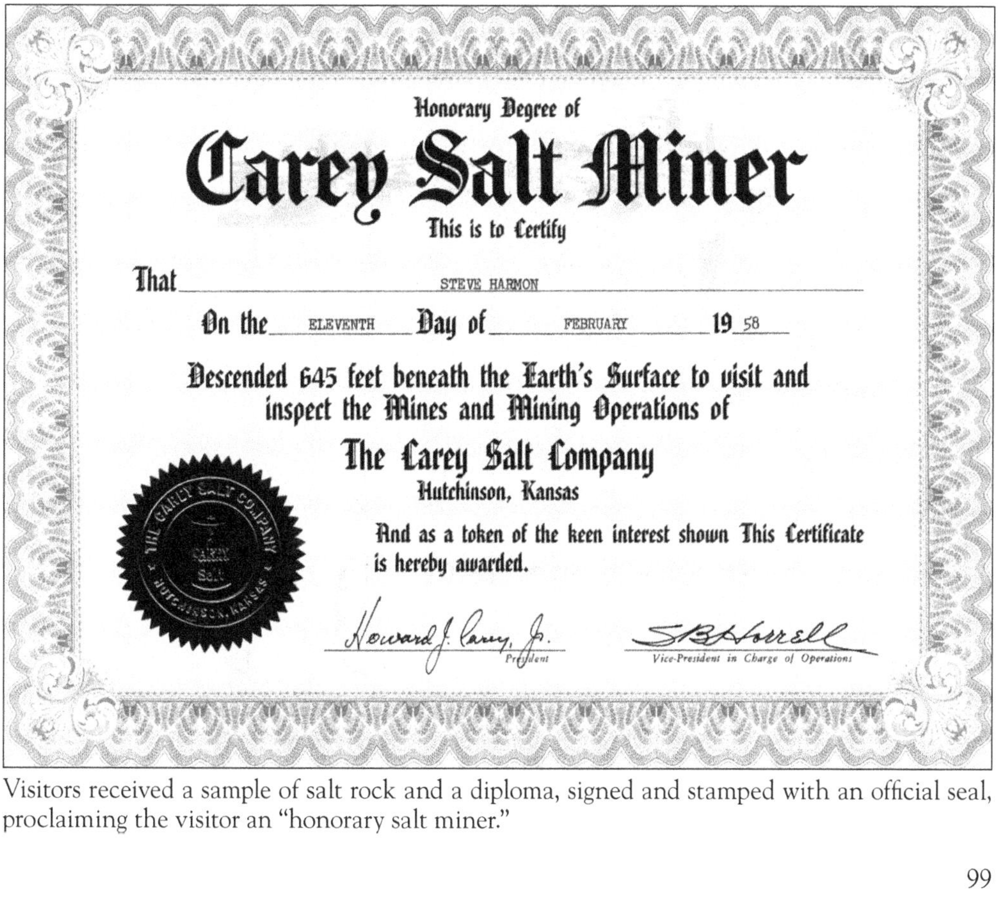

Honorary Degree of

Carey Salt Miner

This is to Certify

That STEVE HARMON

On the ELEVENTH Day of FEBRUARY 19 58

Descended 645 feet beneath the Earth's Surface to visit and inspect the Mines and Mining Operations of

The Carey Salt Company

Hutchinson, Kansas

And as a token of the keen interest shown This Certificate is hereby awarded.

THE CAREY SALT COMPANY · CAREY SALT · HUTCHINSON, KANSAS

Howard J. Carey, Jr. — President

S.B. Horrell — Vice-President in Charge of Operations

Visitors received a sample of salt rock and a diploma, signed and stamped with an official seal, proclaiming the visitor an "honorary salt miner."

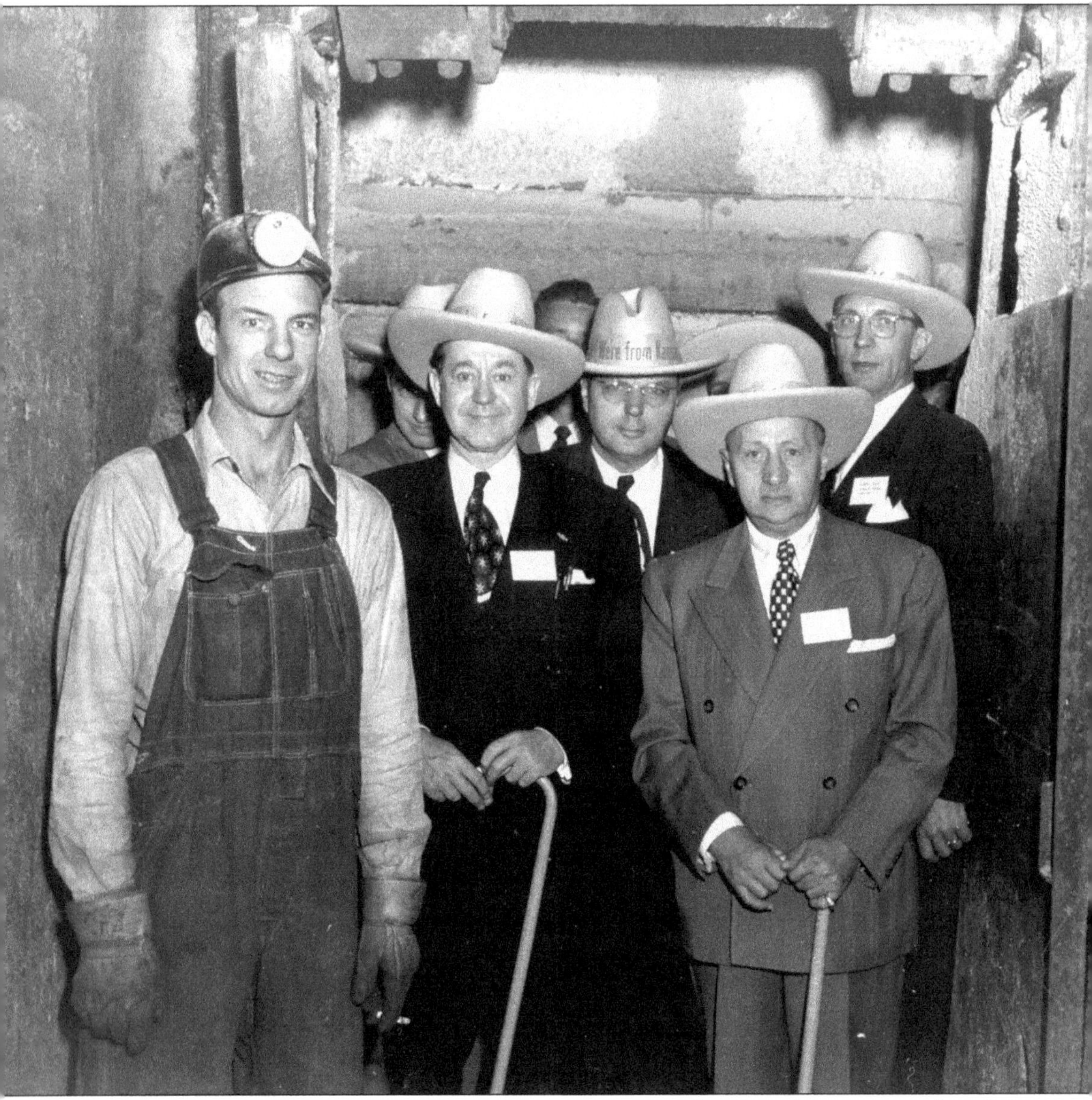

This photograph shows a few members of the Kansas City Chamber of Commerce who, while visiting the 1951 Kansas State Fair in Hutchinson, came to tour the mine. The unidentified signalman poses with them as they stand in the man-cage under the skip.

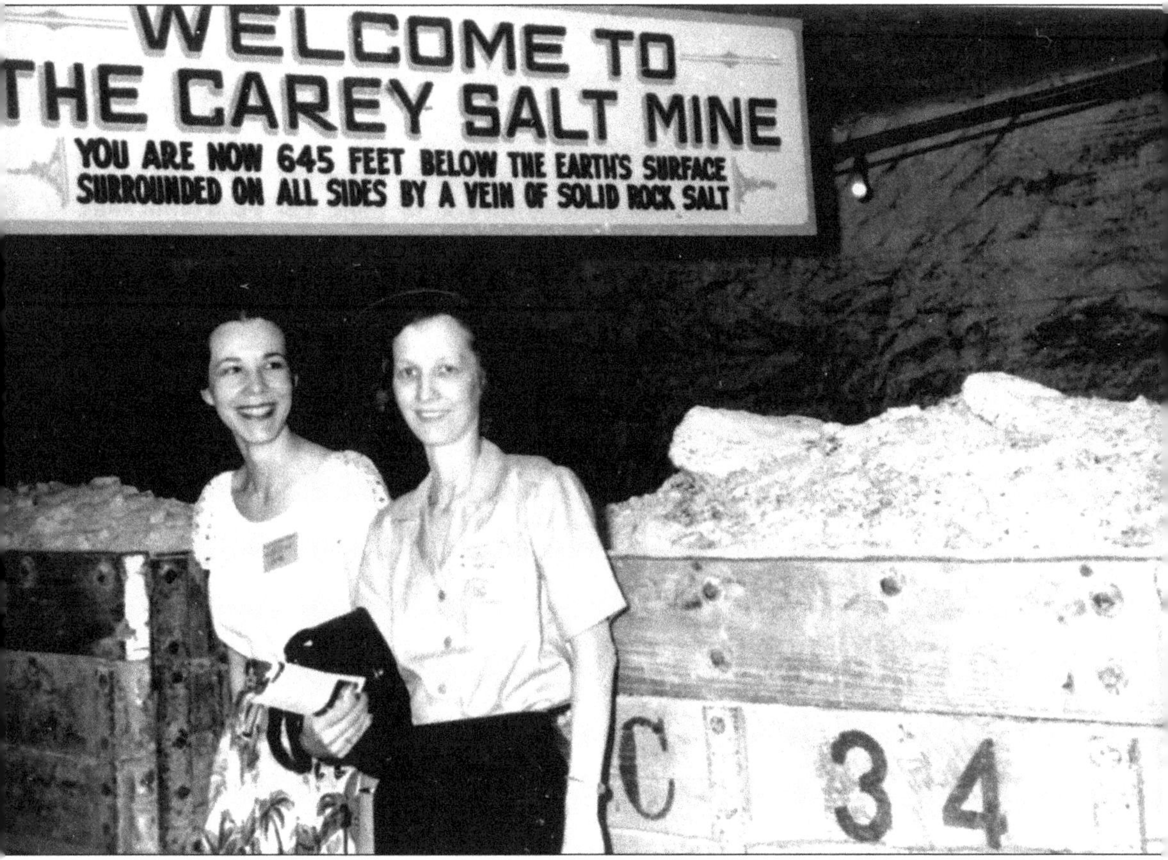

Although during World War II the Carey Salt Company discontinued giving tours, it began taking people underground shortly afterward. A 1951 board of directors report states that demand for salt was so great, "it has been found necessary to cut visiting days at this plant from four days to two." The writer also predicted that it might be necessary to stop giving tours altogether. This was likely wishful thinking, since the very next year brought nearly 9,000 visitors to the mine.

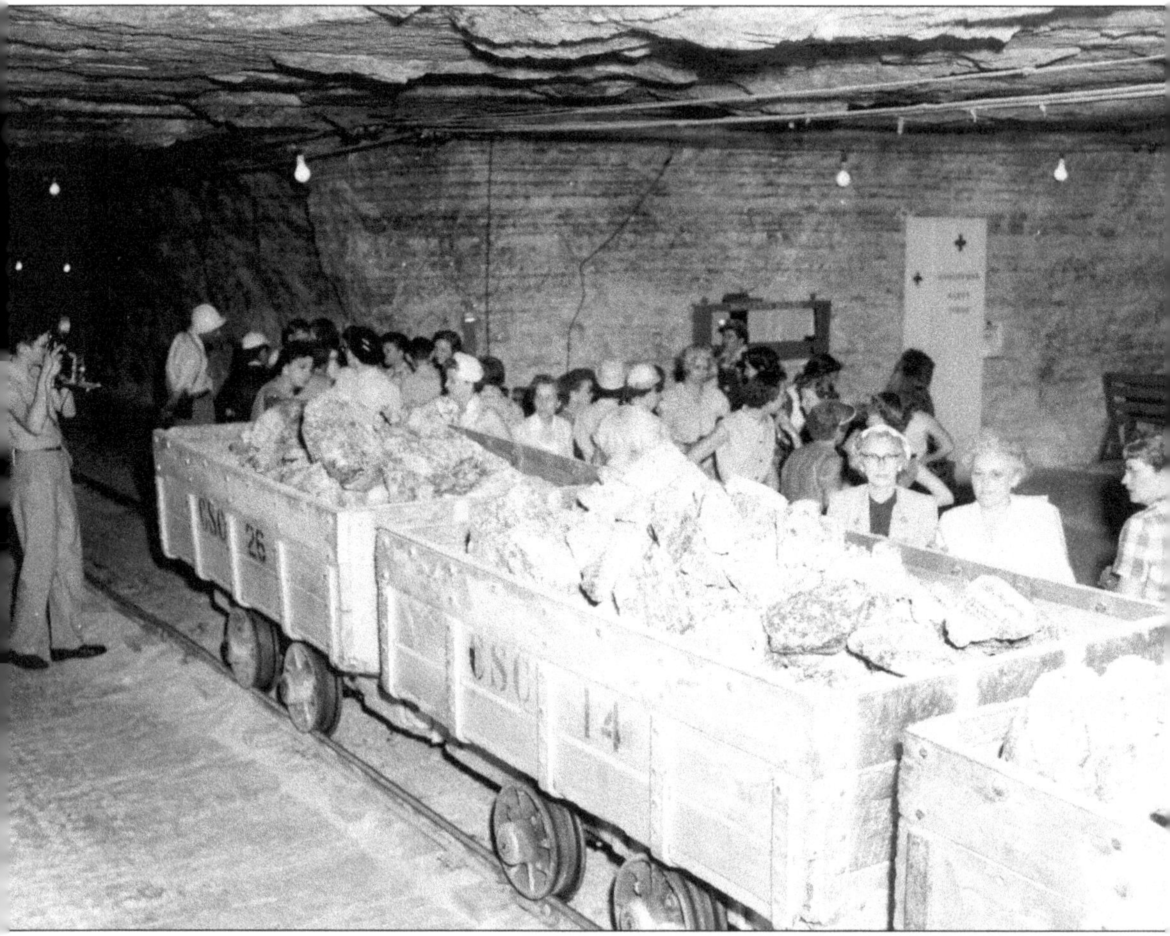

"Due to expensive delays caused by large groups of visitors coming unannounced, a policy of taking visitors who have made arrangements in advance only, has been put into effect. In this way we are able to limit the groups, and, in turn, handle them with a minimum amount of interruption in production." This excerpt from a report to the board of directors came at the end of 1953. Visitors doubtless were a large source of frustration for the miners and supervisors underground, who were trying to get their work done in a potentially dangerous environment. Throughout the 1950s, various compromises were tried, with designated visiting days and policies of only allowing large groups who gave advance notice. The policy of giving advance notice worked better for the miners, who, according to the board of directors reports, continued their work uninterrupted. But in 1958, the policy expanded the number of people allowed in the groups because potential visitors were complaining that the waiting list was more than two years long.

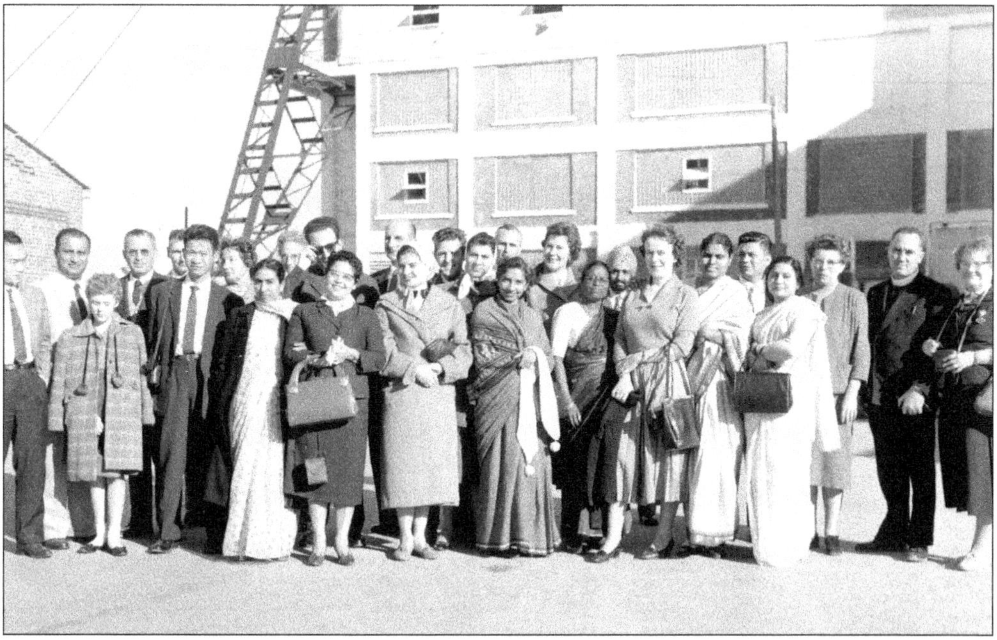

In this 1958 photograph, a group of traveling Fulbright scholars poses before heading underground. Many dignitaries and celebrities toured the mine, including Arthur Fiedler of the Boston Pops, navy officers from the Hutchinson Naval Air Station, and the Phillips Oilers basketball team.

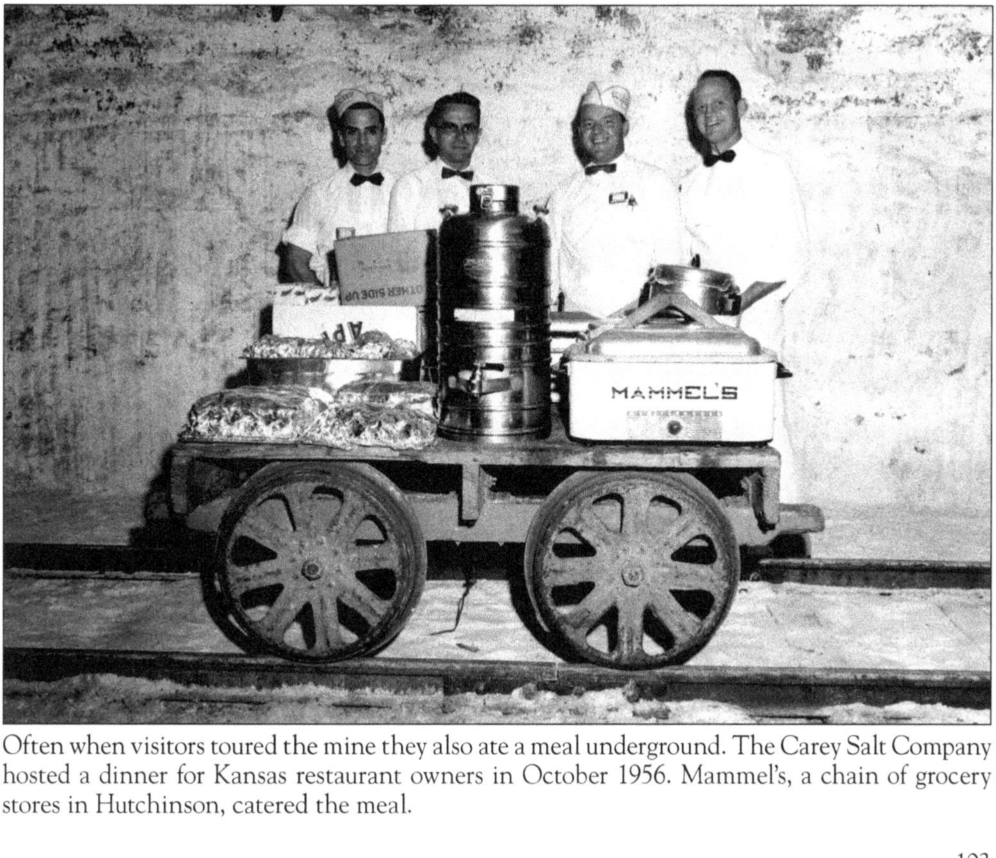

Often when visitors toured the mine they also ate a meal underground. The Carey Salt Company hosted a dinner for Kansas restaurant owners in October 1956. Mammel's, a chain of grocery stores in Hutchinson, catered the meal.

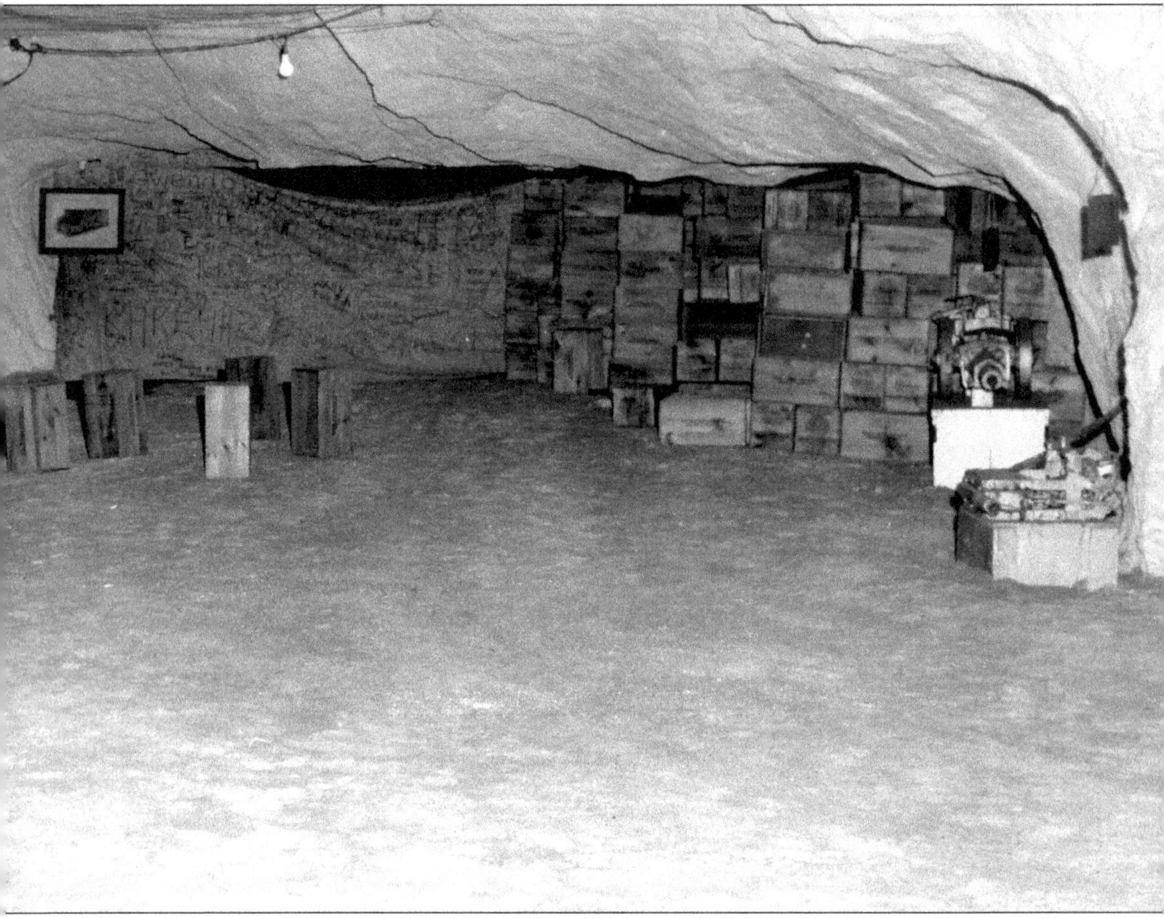

The underground banquet room supplied dynamite boxes upon which visitors could perch while eating their meal. Over the years, many guests signed the brattice cloth in the background. The brattice cloth is used underground to channel the flow of air.

Nine

OTHER USES FOR THE MINE

World War II shaped the future of the Carey Salt Mine in ways that could not be predicted during the conflict itself. When Adolf Hitler invaded Europe, he planned to display the West's great historical treasures that his armies had looted. To protect the pieces until Germany's triumphant victory, the Nazis secreted them away in the Austrian salt mines to keep them safe under the earth's surface. After the Allied forces liberated Europe, American army soldiers went underground and removed the treasures from the salt. One man, Mark Adams Sr., who had heard about this mission, returned home to Wichita shortly after the war.

Nearly 15 years later, the stories of Hitler's dream were still with Adams, and, with his colleague Jack Heathman, he contacted Howard J. "Jake" Carey Jr. with an idea to store records in the unused caverns of the salt mine. Underground Vaults and Storage (UV&S) was born. The new business subleased space from the Carey Salt Company, occupying bays closest to the hoist. Areas that had lain empty after yielding profitable rock salt began once again to make money.

Ironically, widespread fear of nuclear attack helped to make UV&S a success in the first decade, yet it was to find itself a neighbor of the Atomic Energy Commission (AEC). Throughout the 1950s, the AEC had been looking for a safe place to store its growing collection of toxic waste. In 1958, scientists moved into the Hutchinson mine and set up shop. Throughout the next 10 years, they would perform tests in Carey's mines to learn whether salt would be a suitable buffer. By the end of the testing, salt was looking promising. The encouraged AEC announced plans to build a nuclear waste storage facility at the Lyons mine. It met so much political and scientific opposition that in 1972 the AEC left Kansas and headed to New Mexico to build the new facility.

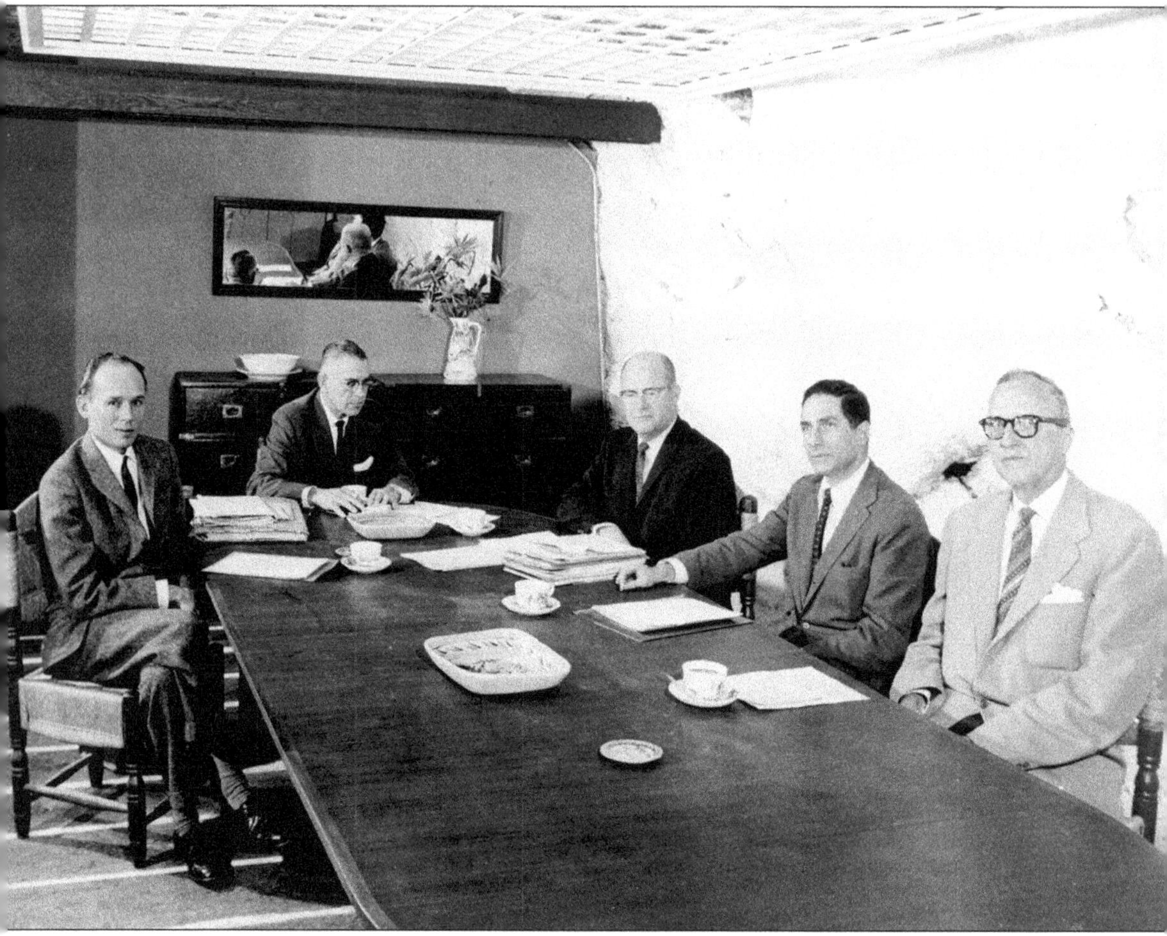

UV&S was incorporated on June 11, 1959, by seven men. From left to right, Robert L. Williams, John Schul (president), J. T. Koelling, S. O. Beren, and Mark Adams Sr. appear around the table in the underground conference room in this 1960 photograph. Not pictured are directors Jake Carey, Jack H. Heathman, and Joe J. Honaker. Originally the founders of the company were the only stockholders in the corporation, which authorized 1,000 shares of common stock at a par value of $100. (Courtesy UV&S.)

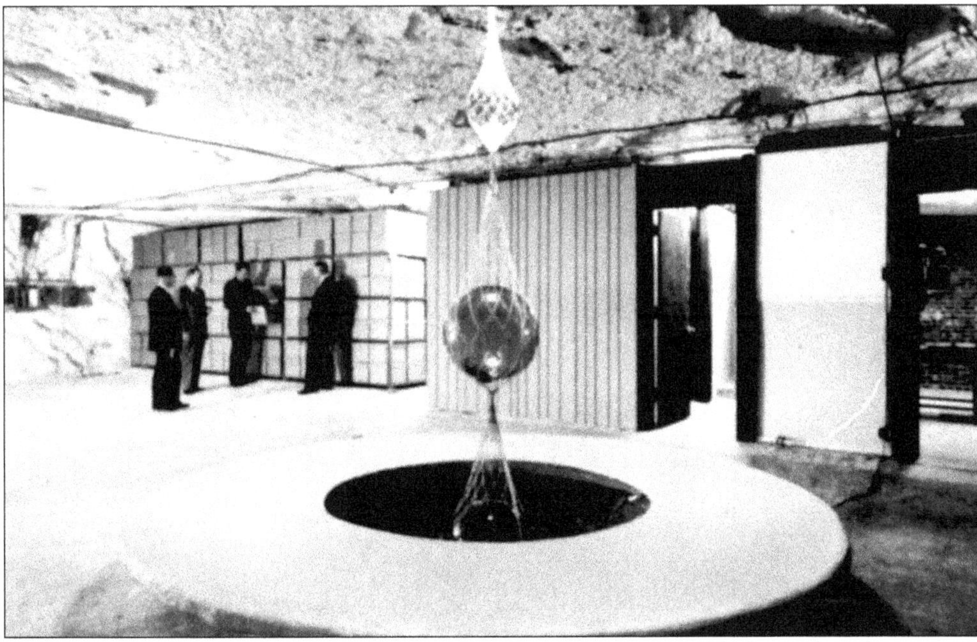

In the early years, this fountain in the foyer, which was home to 13 goldfish, marked the entrance to UV&S's offices. The fledgling company could fit its storage and offices in the same bay. UV&S operated at a loss until 1965. (Courtesy UV&S.)

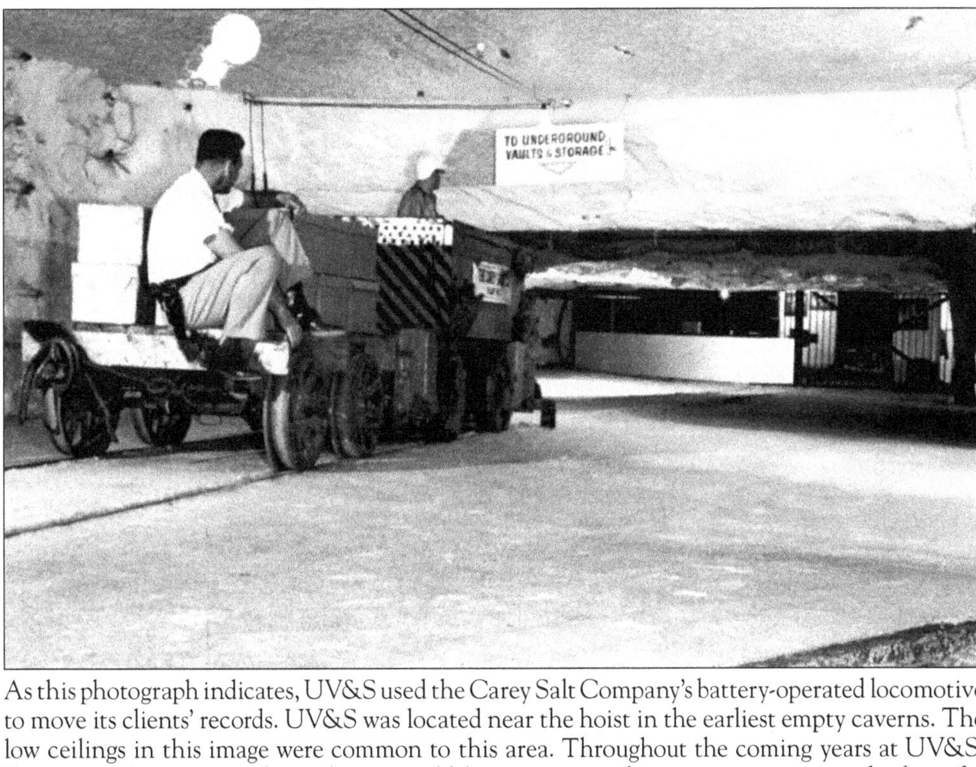

As this photograph indicates, UV&S used the Carey Salt Company's battery-operated locomotive to move its clients' records. UV&S was located near the hoist in the earliest empty caverns. The low ceilings in this image were common to this area. Throughout the coming years at UV&S, "hi-topping," or raising the ceilings, would be a common first step in preparing the bays for storage. (Courtesy UV&S.)

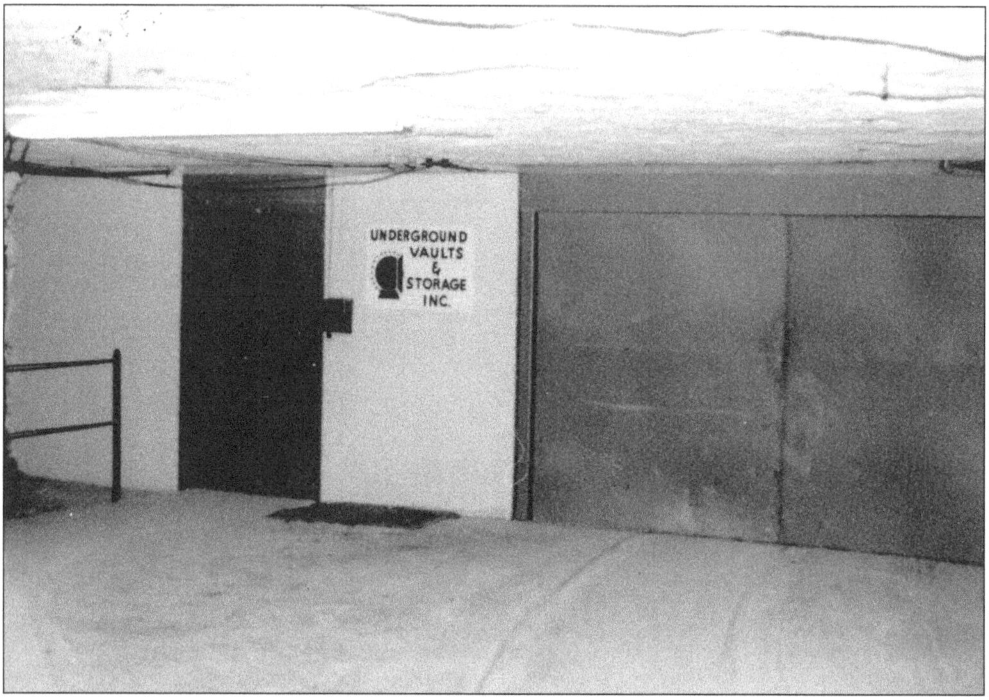

In later years, the entrance was changed to a solid wall instead of steel bars. This photograph dates to the late 1970s.

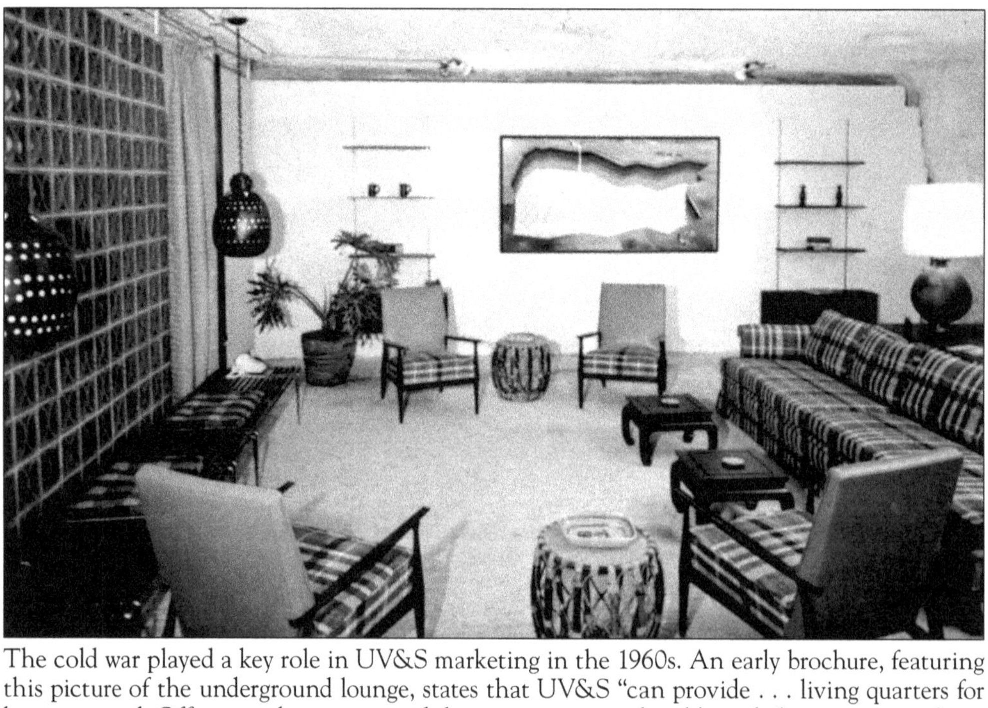

The cold war played a key role in UV&S marketing in the 1960s. An early brochure, featuring this picture of the underground lounge, states that UV&S "can provide . . . living quarters for key personnel. Offices can be constructed that are just as comfortable and pleasant as any above ground, so that a company could move its basic staff to safety and maintain its functioning without interruption." (Courtesy UV&S.)

When Leora Hard, pictured, began at UV&S in 1965, she was one of seven employees working underground. Her required uniform was gradually phased out as more employees were added to the staff roster. Starting at $1.25 per hour, she was responsible for many tasks, becoming a room operator for the open-shelf storage and eventually becoming a supervisor. (Courtesy UV&S.)

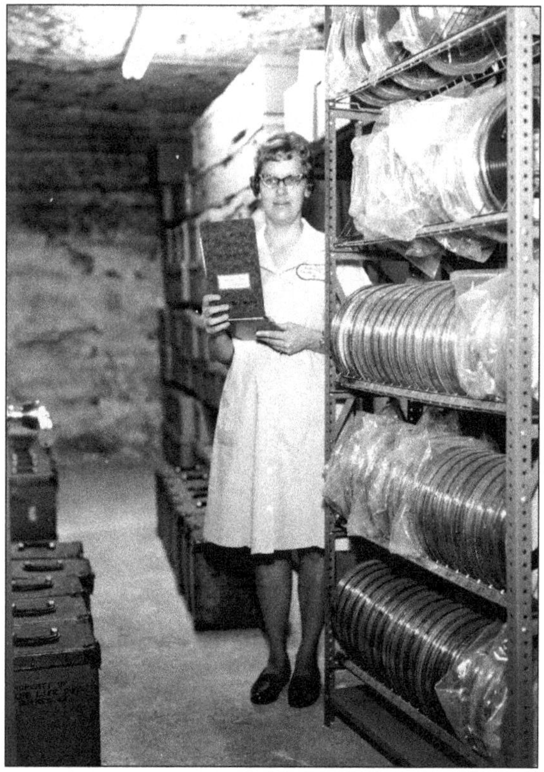

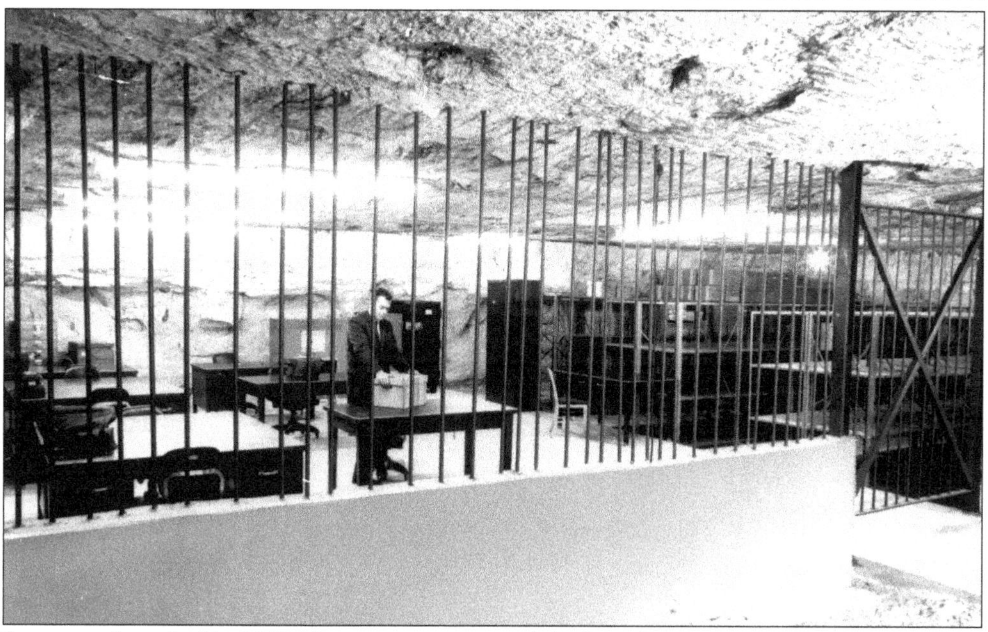

Power was furnished by UV&S's own generators, with backup electricity coming from the Kansas Power and Light Company. The area lights ran on direct current, which was converted from the alternating current coming down the shaft. This image shows the facility of the Arabian Oil Company, which in the early 1960s leased an entire 50-by-300-foot bay for storing its records. (Courtesy UV&S.)

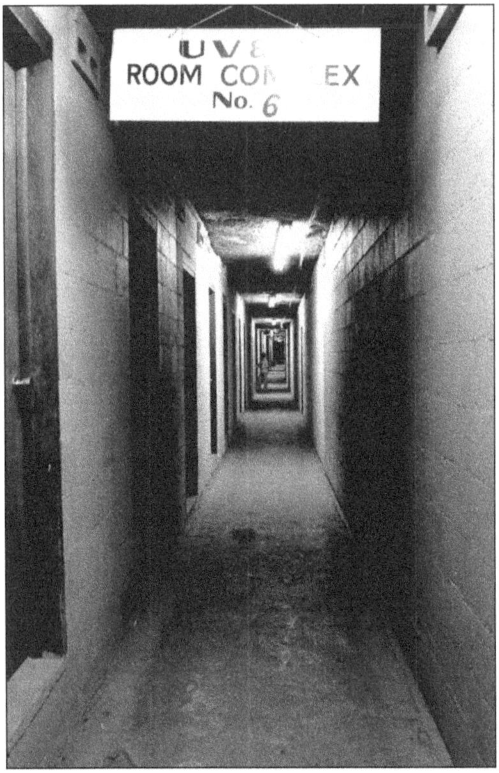

Clients with especially sensitive materials could lease their own rooms within one of the bays. Workers would even install special climate-control equipment when requested by a client. (Courtesy UV&S.)

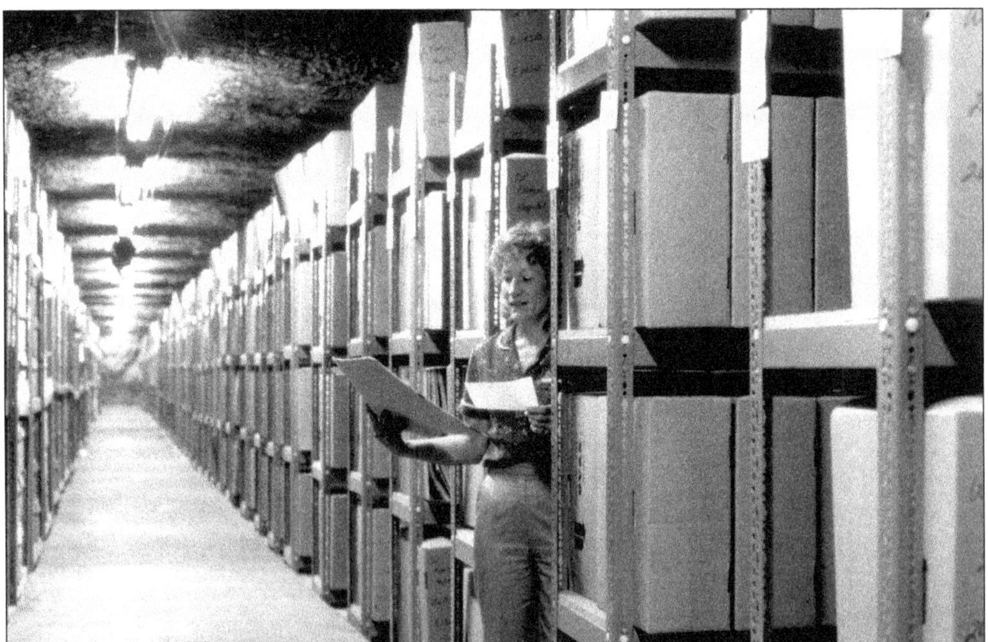

Not all clients opted to store their items in a private room. Most chose the "economy storage," which put their records with the materials of other clients along shelving in huge bays. UV&S has maintained a sophisticated data retrieval system to keep track of each box. In this photograph, UV&S employee Peggy Nikkel prepares to retrieve a box. (Courtesy UV&S.)

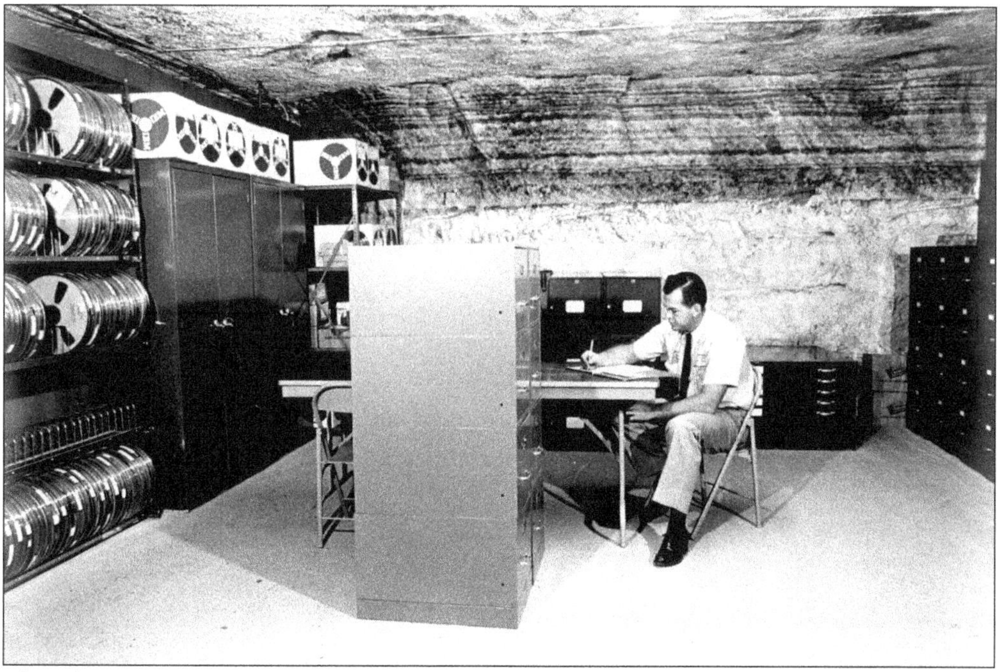

This photograph appeared in a brochure showing the different types of space available in the 1960s. The "type II private room" totaled 420 square feet and could be "equipped for storage of all types of essential records, i.e. magnetic tape, microfilm, engineering drawings, original documents, etc." Magnetic media was a mainstay of storage materials underground, as was microfilm and fiche. (Courtesy UV&S.)

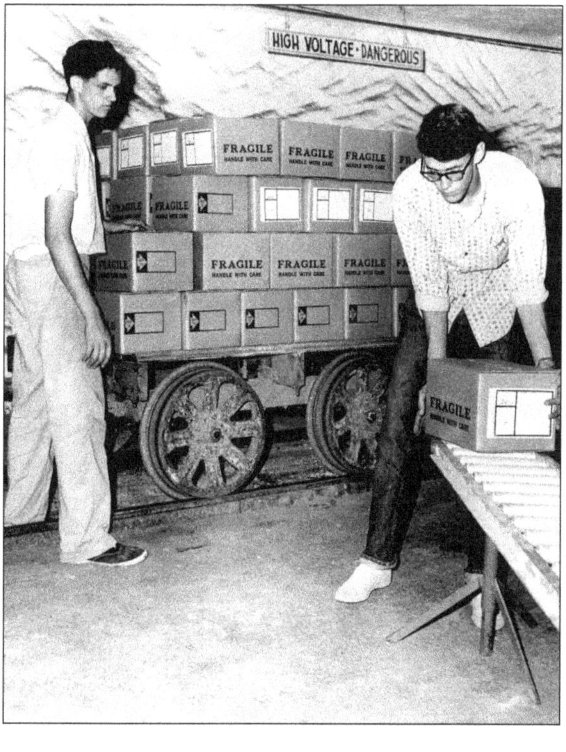

At UV&S, the railroad tracks from the hoist only went so far. Once they ended, workers loaded the freight manually onto pushcarts and hand trucks to get them to their final destination in the storage bays. Later electric stock chasers lightened the job considerably. (Courtesy UV&S.)

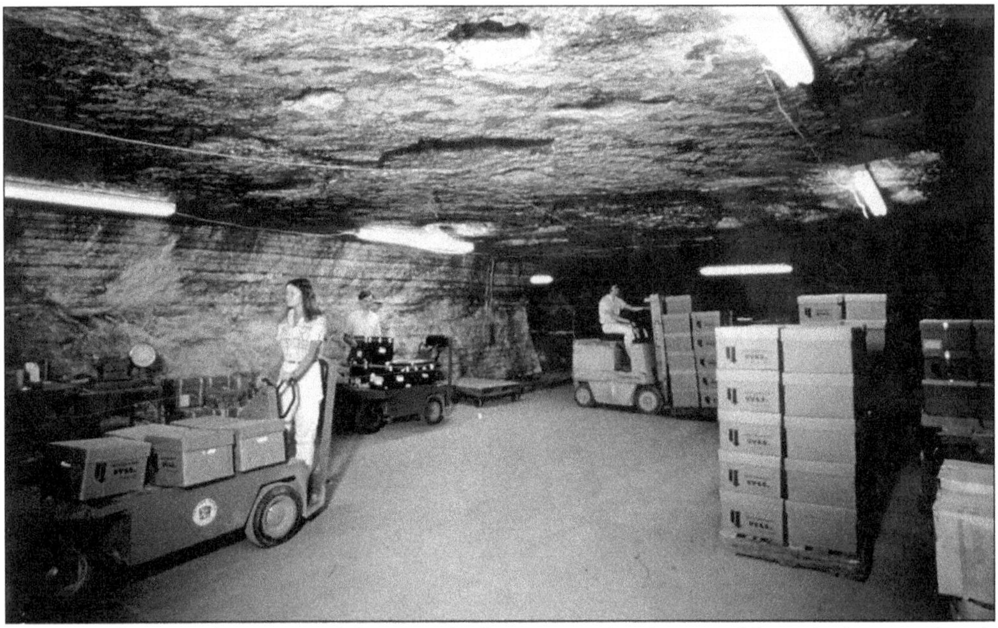

Employees have used a variety of electric vehicles to move freight to and from the hoist and work spaces in the last two decades, including forklifts, stock chasers, and hand trucks. Employees moved themselves on golf carts or tricycles to cover the vast amount of space in the work and storage areas. (Courtesy UV&S.)

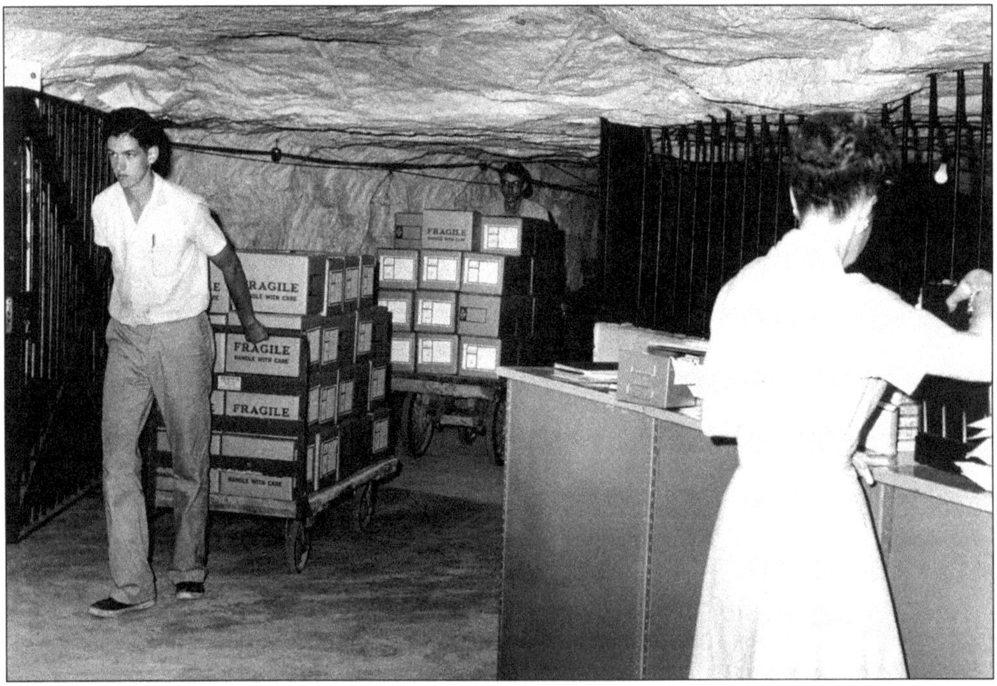

Transferring large volumes of records across the country could be a nerve-racking experience for the clients. Two executives from one company flew to Hutchinson to supervise the unloading of 600 boxes of microfilm in the 1960s. They used college students, shown here bringing the boxes into the facility, to do the bulk of the work. (Courtesy UV&S.)

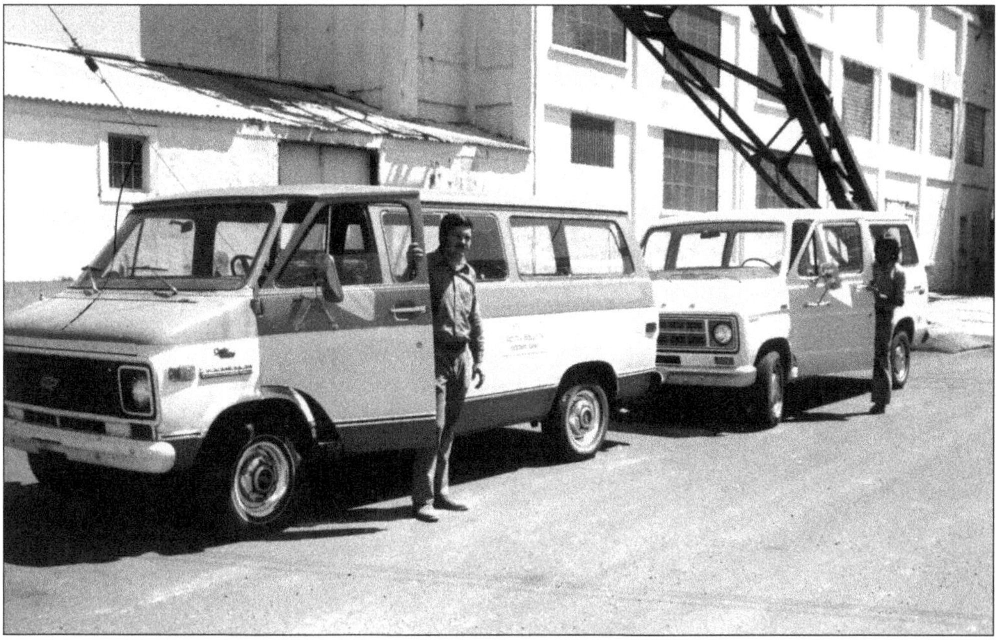

At some point, UV&S began to offer twice-daily pickups from the post office, railway express office, and Hutchinson Municipal Airport. The drivers of these vans stand next to the shaft, ready to transport clients' records in this late-1970s photograph. Eventually 24-hour retrieval service, seven days a week, was implemented. (Courtesy UV&S.)

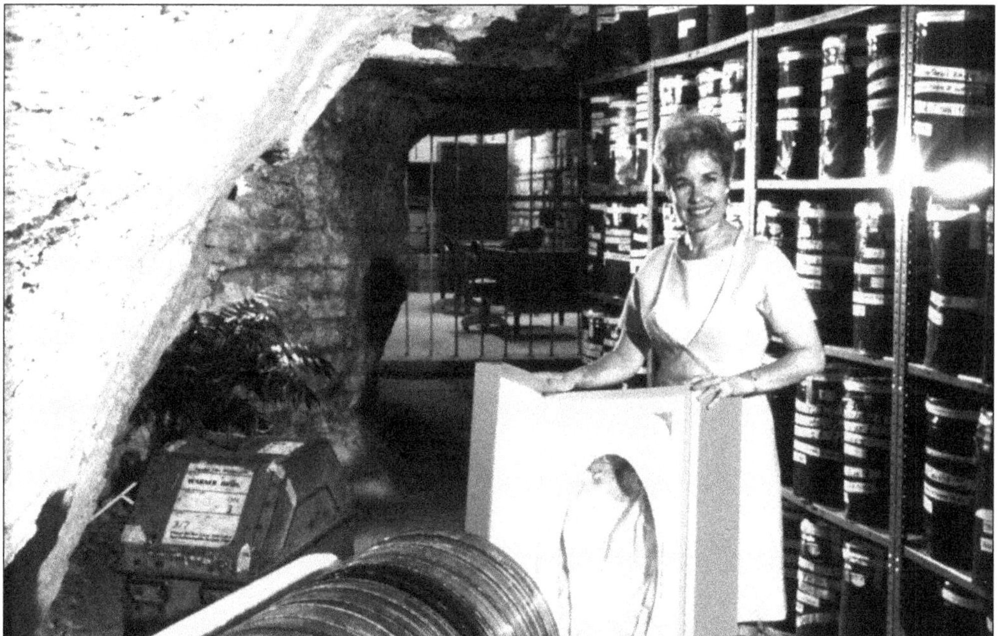

This promotional photograph shows an unidentified UV&S employee showcasing the many different items that the company was storing underground, including a wedding dress, maps, and films. UV&S did not take long to tap into the market of film storage, and today millions of reels are kept underground from several different movie and television companies.

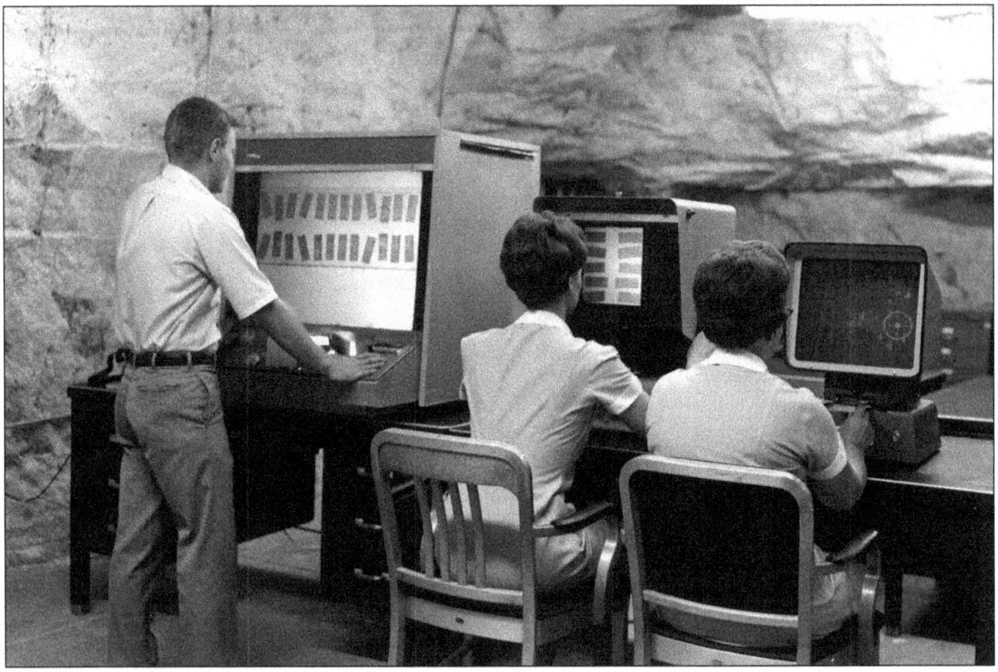

UV&S offered more than merely storage space. UV&S operated a full microfilming laboratory in which it could make microfilm negatives, prints, and enlargements for its clients. Retrieving the data stored underground for its clients has remained a major component of the business. (Courtesy UV&S.)

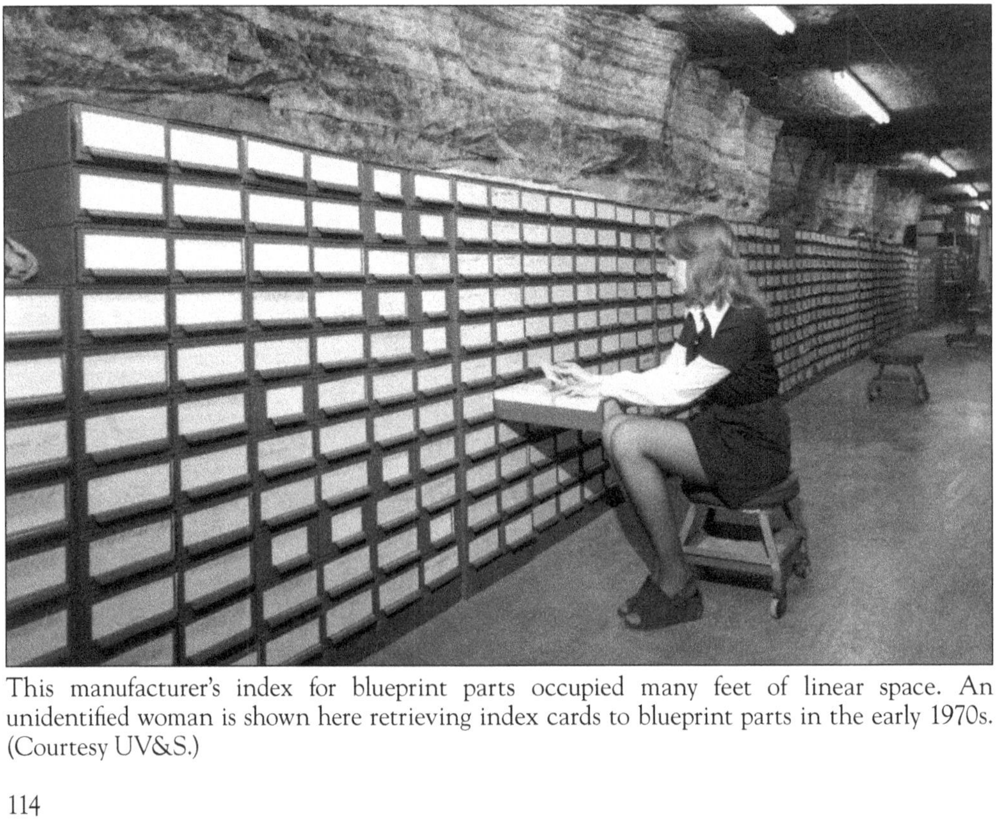

This manufacturer's index for blueprint parts occupied many feet of linear space. An unidentified woman is shown here retrieving index cards to blueprint parts in the early 1970s. (Courtesy UV&S.)

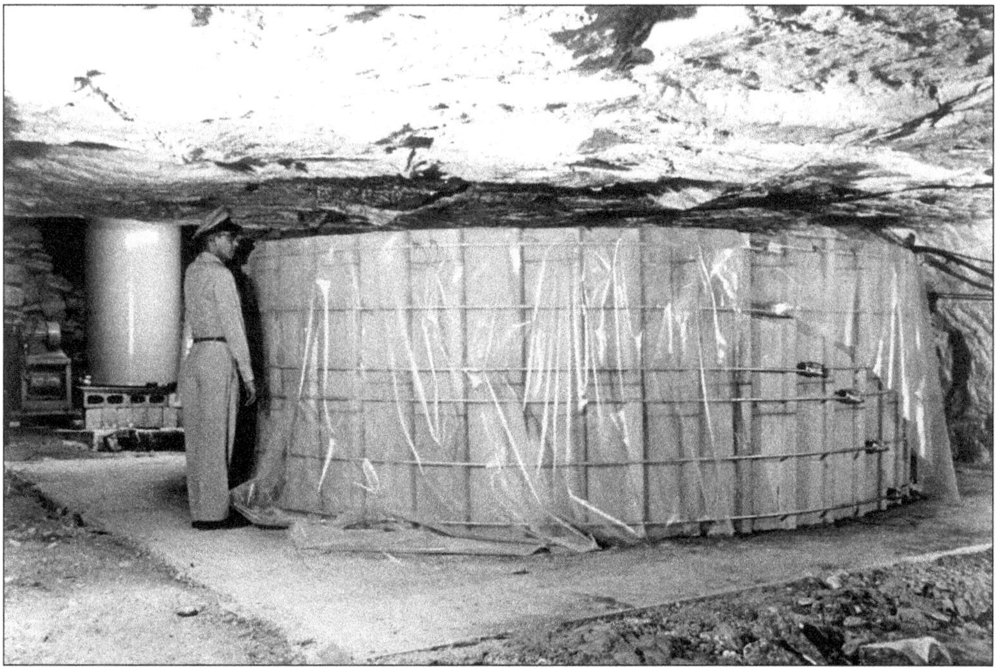

Paving the storage bays required water, as did the growing number of employees. Shortly after the company formed, UV&S installed a 70,000-gallon water tank underground. Over the years, UV&S employees had to paint the tank's interior. The worker could not be very large because the only access to the inside was between the low salt ceiling and the top of the tank. (Courtesy UV&S.)

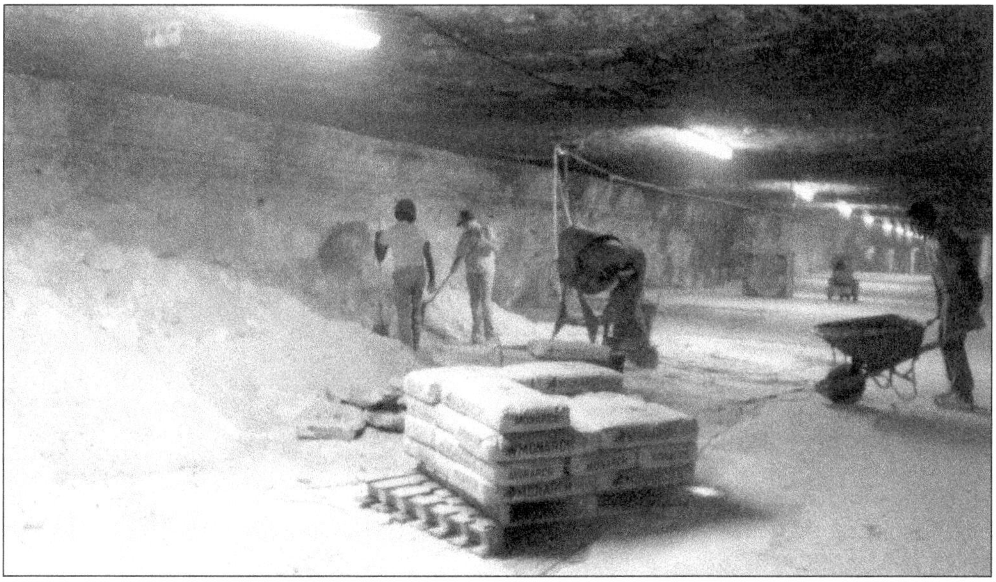

Before the space could be leased, it first had to be prepared. First workers strung lights. The workers in this 1986 photograph are about to pave the floor in sections with saltcrete—a combination of cement, water, and salt fines. Afterward these workers would have sealed the floor and painted the ceilings and walls white to increase the light and decrease dust. (Courtesy UV&S.)

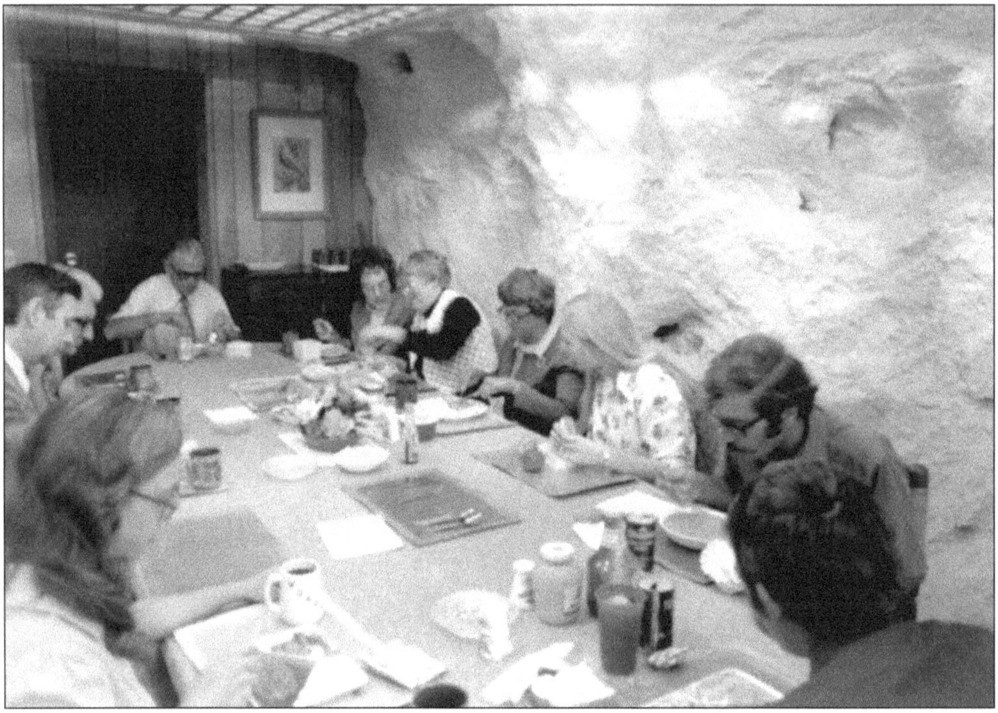

This image from around 1980 shows employees eating lunch. Perhaps because the employees cannot leave the area for lunch and breaks, a strong sense of community exists among them. At the head of the table, facing the camera, is John Schul, who was president of UV&S from 1960 to 1983. He was the first employee hired at the company. (Courtesy UV&S.)

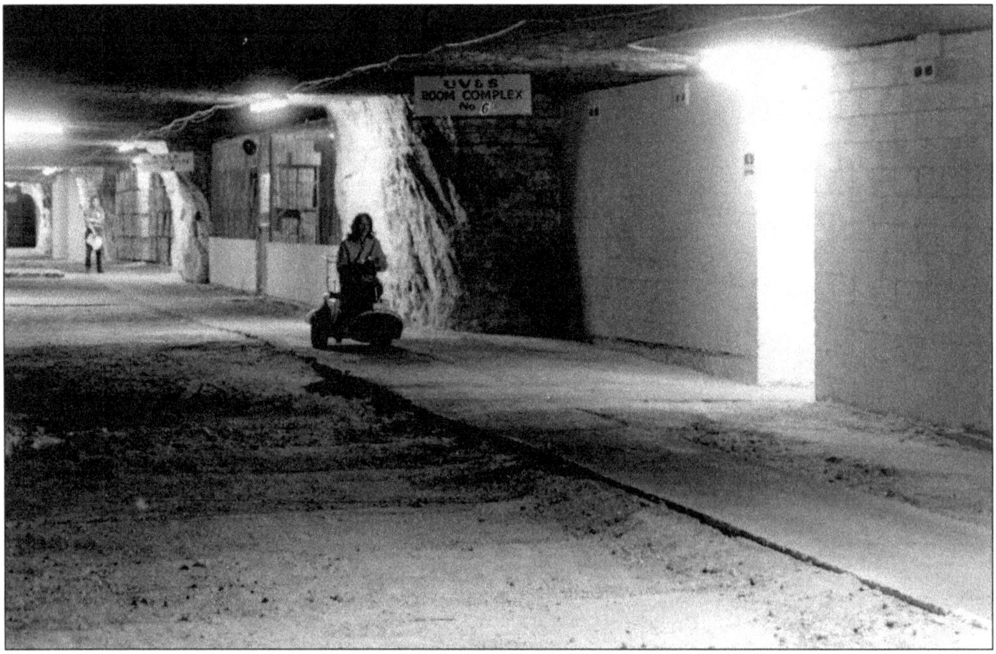

In 1999, 40 years after the company's incorporation, UV&S decided to expand its storage facility by 400,000 square feet, bringing the total storage underground to 38 acres. (Courtesy UV&S.)

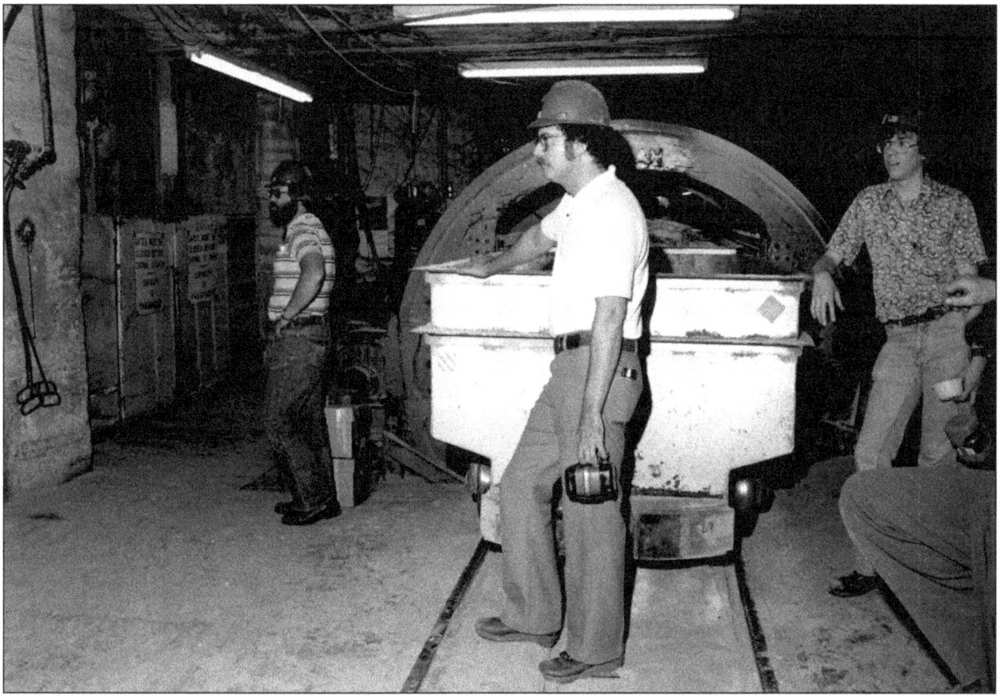

For the first several decades, the UV&S employees rode the salt skips with the miners until late 2006, when a new shaft was opened southeast of the original hoist. The new hoist enabled UV&S to run freight and employees without halting the production of salt. This late-1970s photograph shows UV&S employees waiting for the hoist to take them topside at the end of their shift. (Courtesy UV&S.)

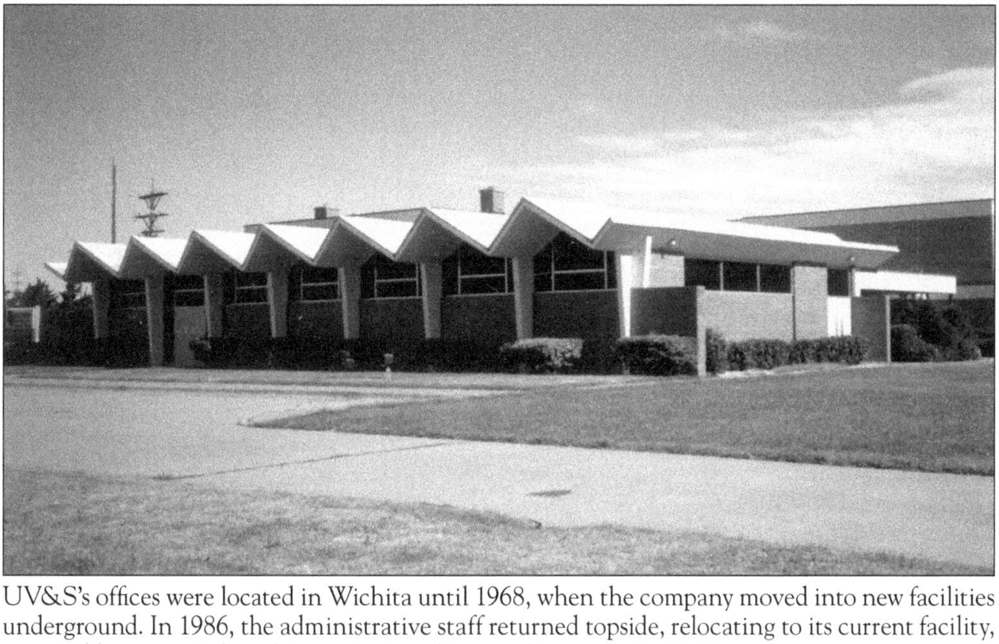

UV&S's offices were located in Wichita until 1968, when the company moved into new facilities underground. In 1986, the administrative staff returned topside, relocating to its current facility, pictured here, on Halstead Street in Hutchinson. (Courtesy UV&S.)

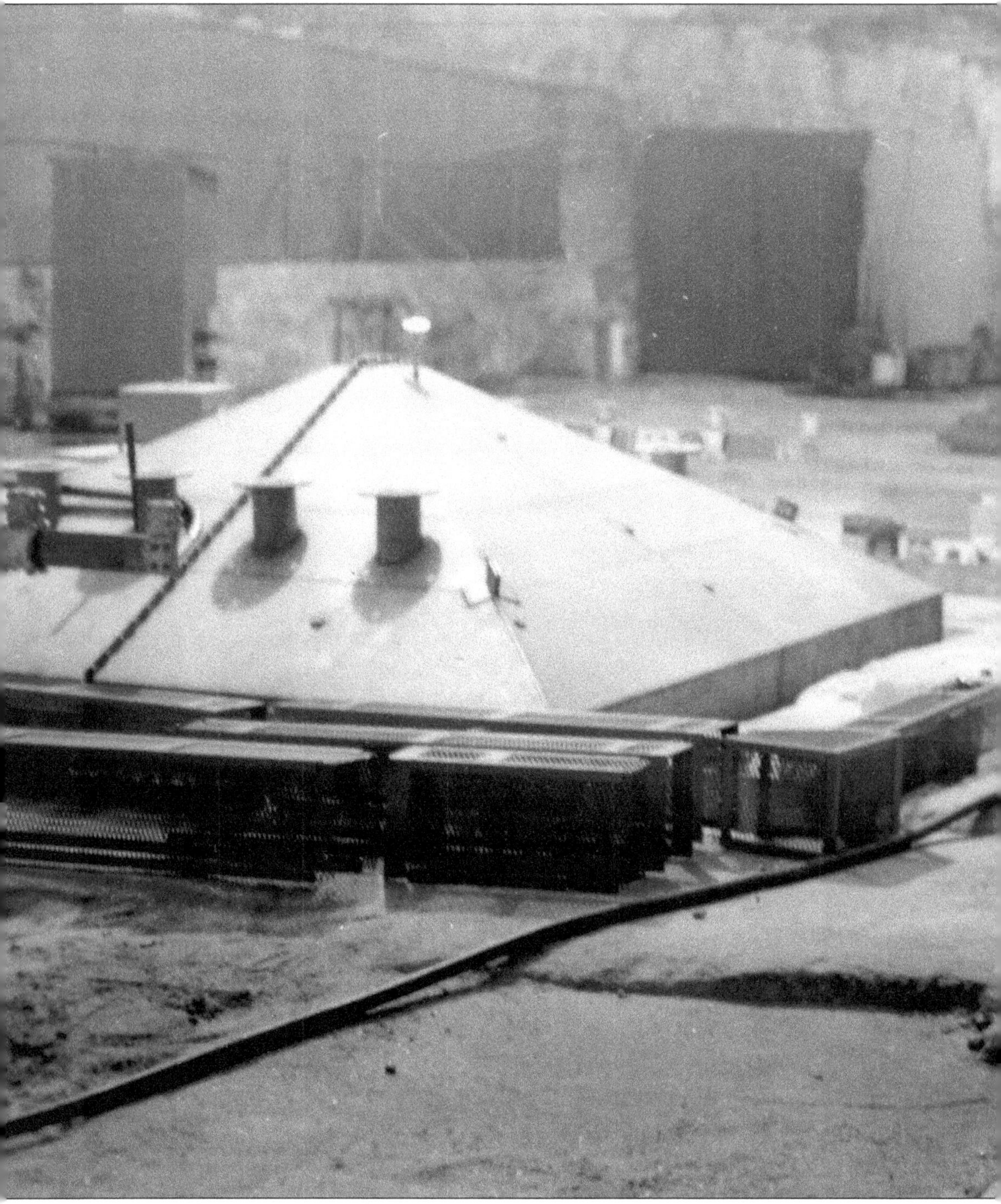

The December 1958 vice president of operations' directors report states, "A tentative estimate was presented by the Carey Salt Company for consideration by the Oak Ridge National Laboratories and approval of the AEC. Our estimate was to be considered along with an estimate of another salt company. We were advised on December 2, 1958 that they had selected the mine here at Hutchinson as the location for the test in question. It has been agreed that the work will be

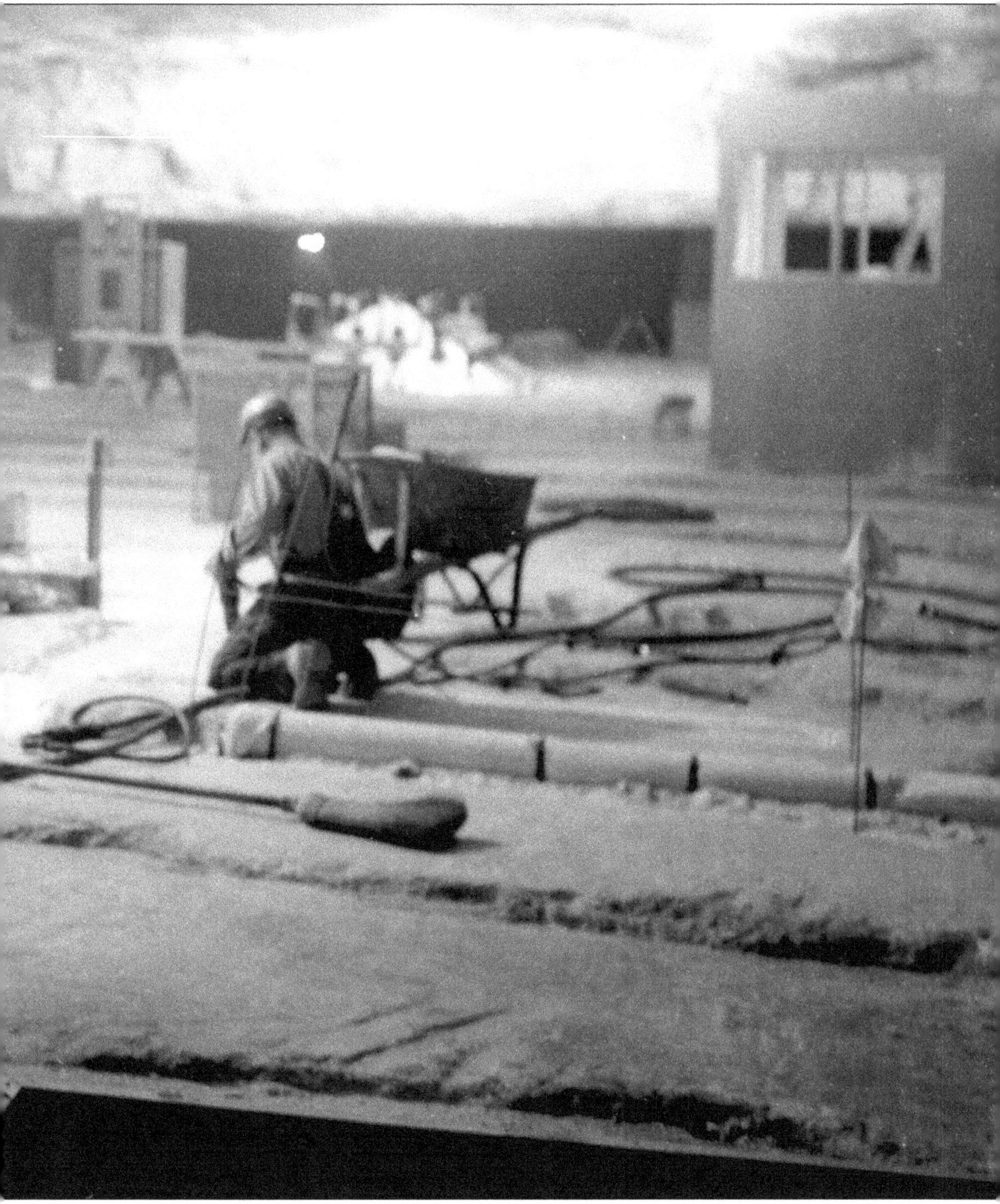

carried on by the Carey Salt Company on a cost plus fixed fee basis, under the direction of the Oak Ridge National Laboratories. . . . It is our understanding that tests will be publicized quite generally. A representative from the public relations department will be on hand to direct this phase before the tests are started. Visitors from all countries of the world are expected to observe the tests from time to time. The test will extend possibly over a two year period."

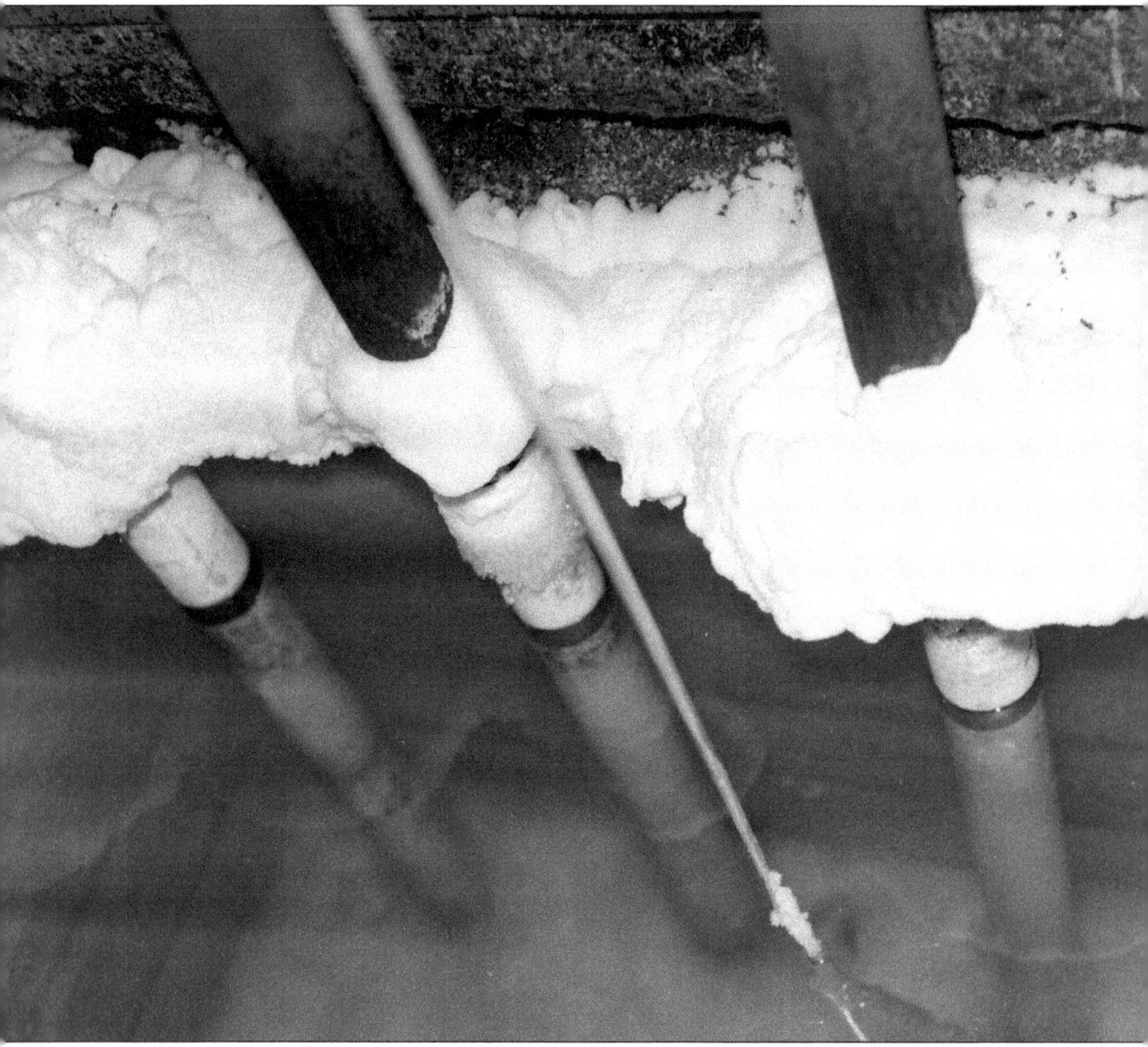

The tests were predicted to last for two years but in reality lasted until 1967. They consisted of a series of experiments with nonnuclear materials to simulate conditions of nuclear storage in the mine. The vice president of operations states in a 1959 report, "The work has been most interesting and we feel that we have learned a great deal about our salt, as well as salt deposits in general, in carrying on the work of this contract. . . . Several meetings of scientists from all parts of the country have been held at the location and some very interesting points discussed."

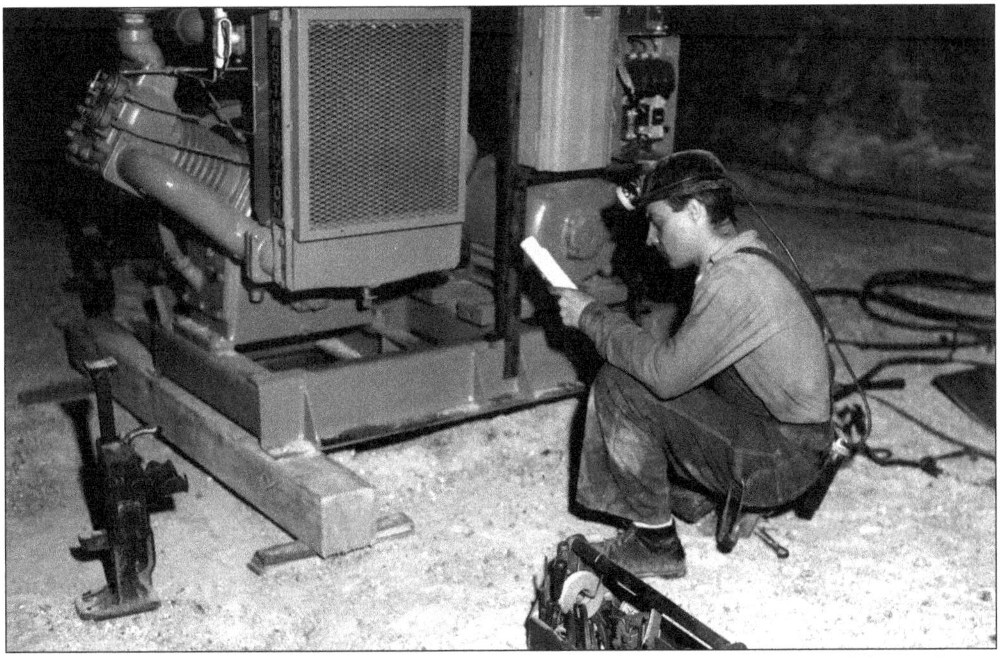

This miner is assembling an air compressor in the Hutchinson mine. Carey Salt employees conducted much of the work underground, including preparing rooms, installing equipment, and monitoring some of the tests.

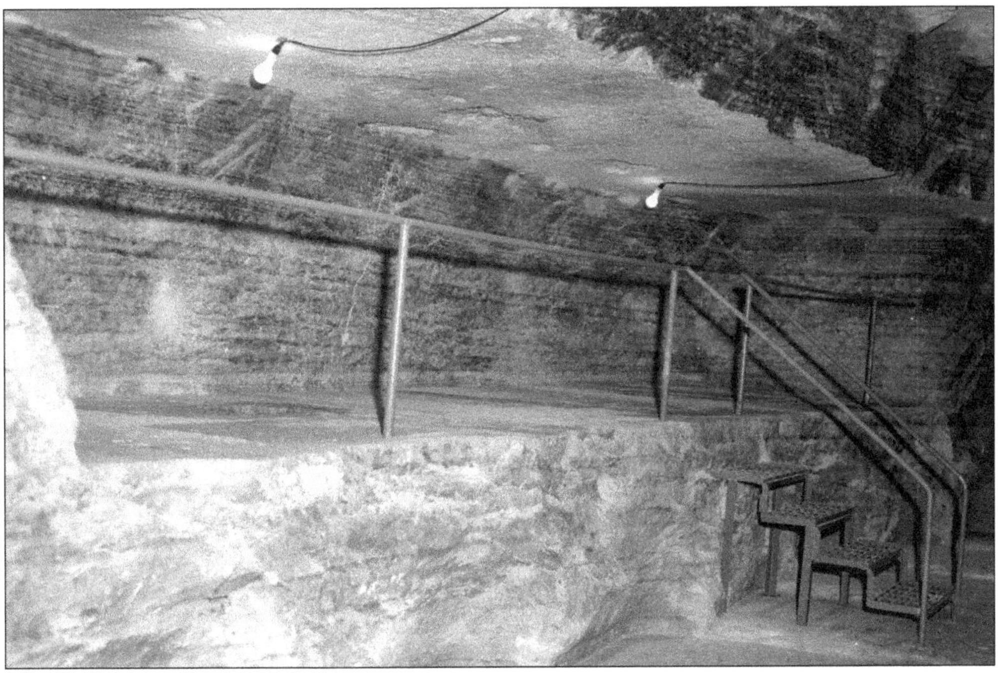

This area, known to Hutchinson Salt miners as "the stage," was cut with an undercutter and blasted out. Carey Salt billed Union Carbide, which had contracted with the AEC, for labor and material costs.

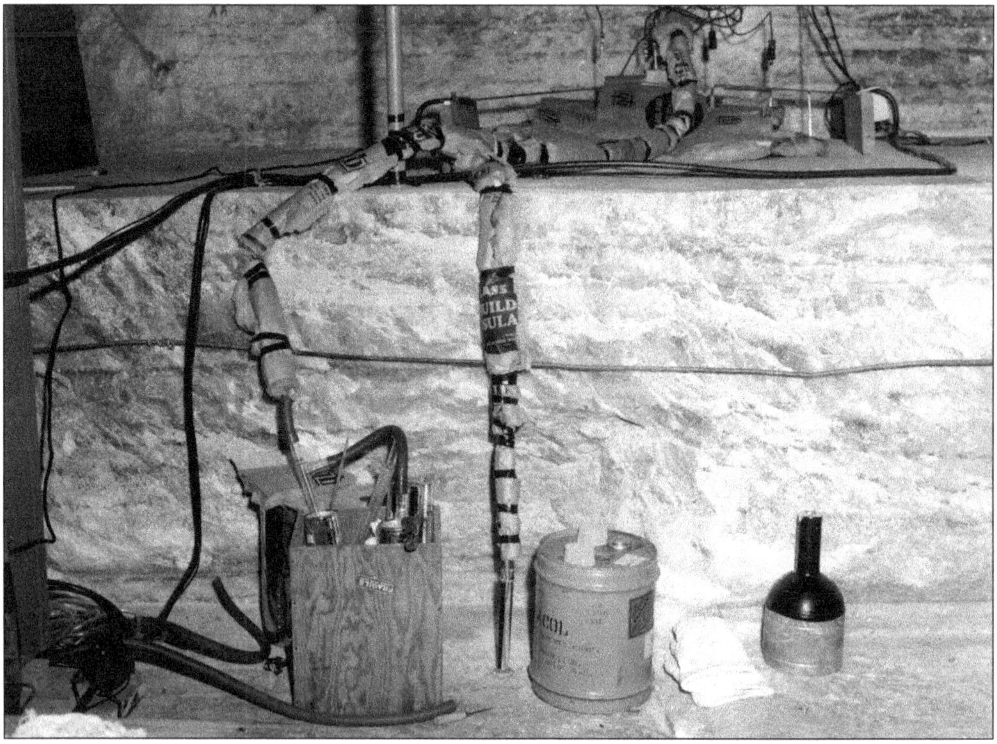

This experiment was located on the surface of and on the floor next to the stage.

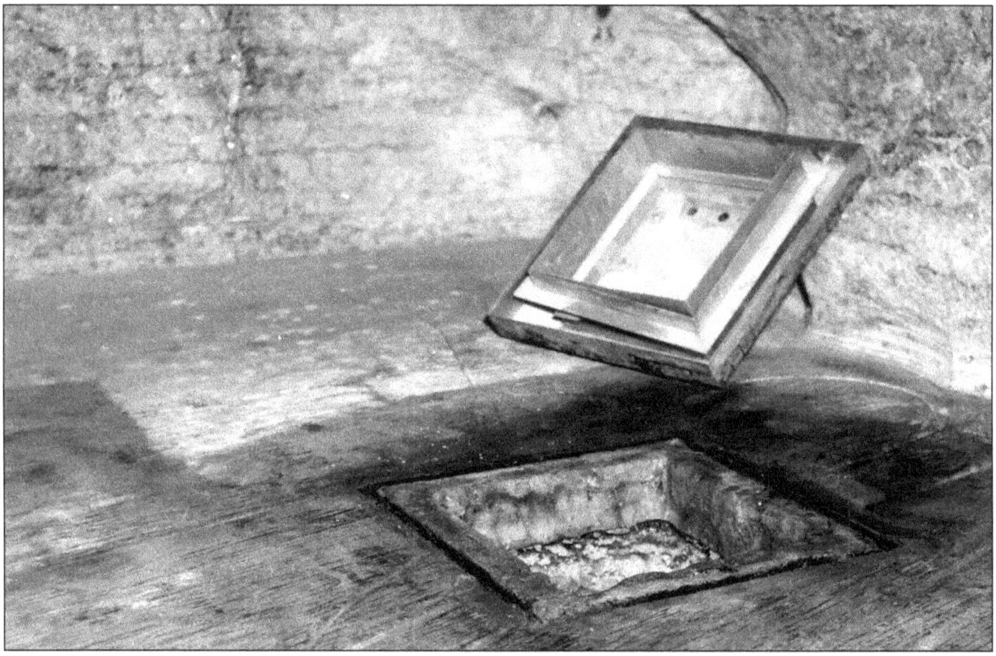

Holes were created and chemicals placed inside to see how the salt would react. Salt, because it is plastic and tends to stretch before breaking, was thought to self-repair any cracks or holes, making it impermeable. The plan was to drop the canisters of toxic waste into holes cut in the salt floor, then fill the rooms with loose salt.

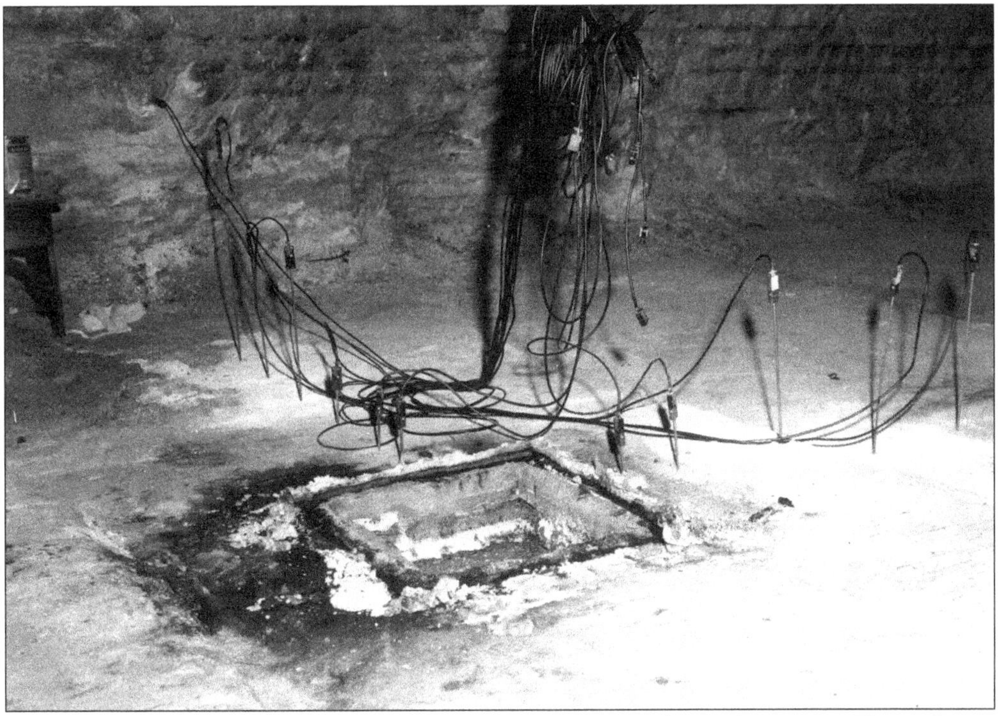

Scientists hoped the salt would help to dissipate the heat from the nuclear waste. The probes in this experiment measured the heat radiating from the point of storage.

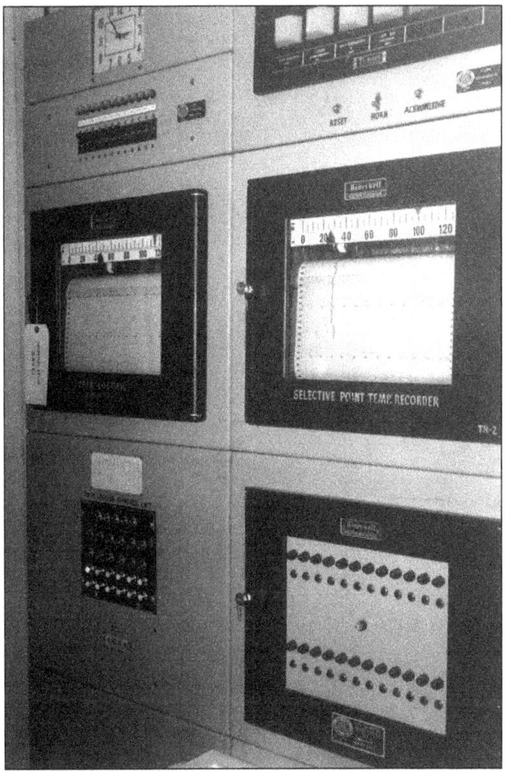

Scientists graphed the test data with these monitors.

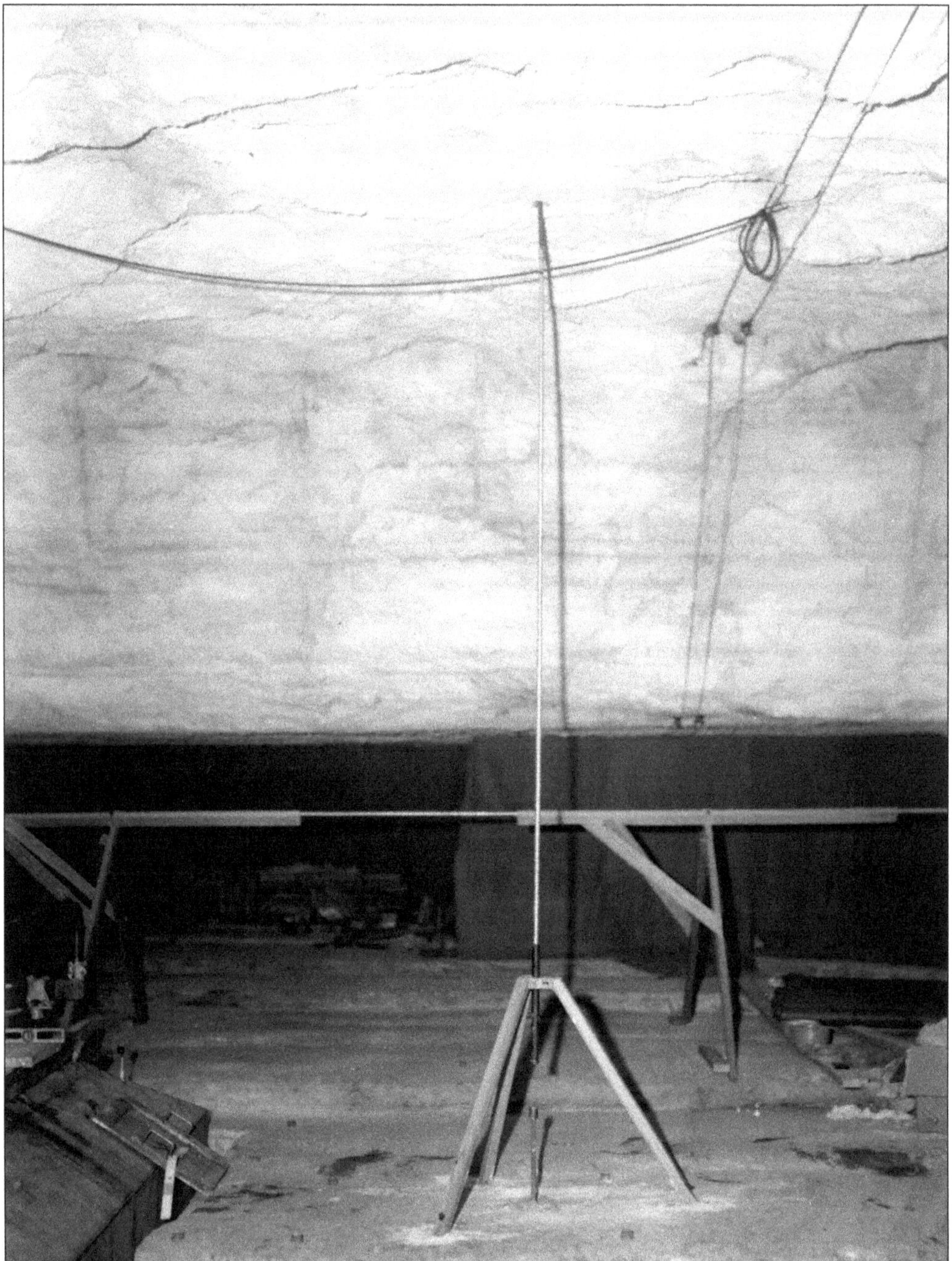

Not only was the AEC interested in the chemical properties of the salt, but they also wanted to know how the caverns held up over time. Devices like this roof monitor were stationed all over the mine to measure how fast the ceiling would begin to sag. Scientists wedged a long, metal pole into the ceiling and inserted a corresponding post in the floor below it. They measured the gap between them periodically to detect any vertical movement. The tripod was placed around the top post to keep it steady. The AEC workers also erected devices with piano wire to detect lateral movement.

The AEC took core samples of the salt bed and surrounding rock formations. Many were concerned that somehow the water table above the salt bed could become irradiated. Public anxiety about the water issue was one of the reasons the AEC finally pulled the project out of Kansas.

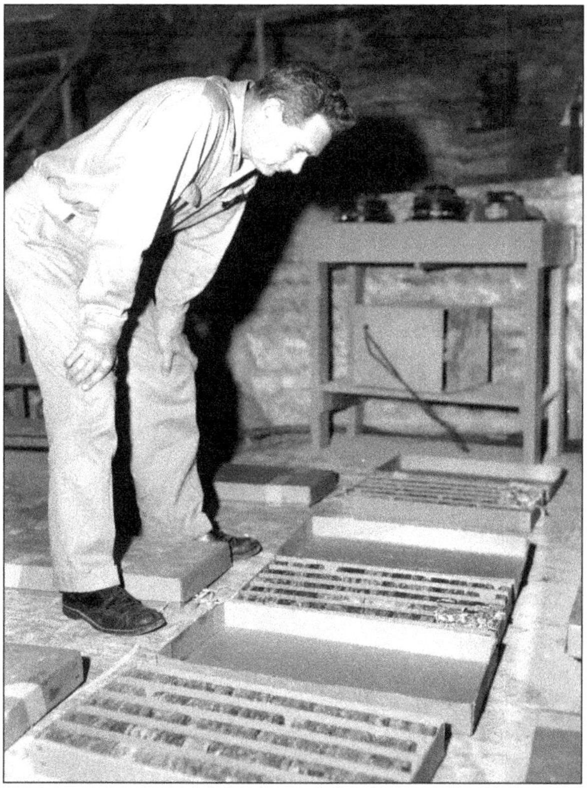

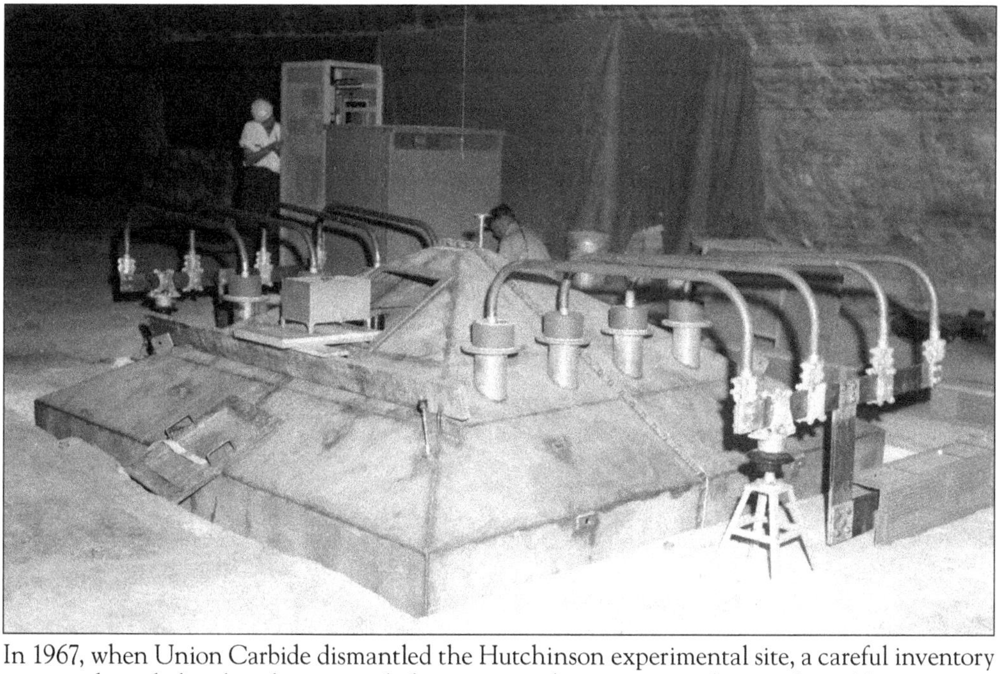

In 1967, when Union Carbide dismantled the Hutchinson experimental site, a careful inventory was conducted that listed items including a control room, a stainless steel scrubbing system, caustic soda, silica gel, a plastic flow system installed by the University of Texas, cooling clamshells, and 37 thermocouplers.

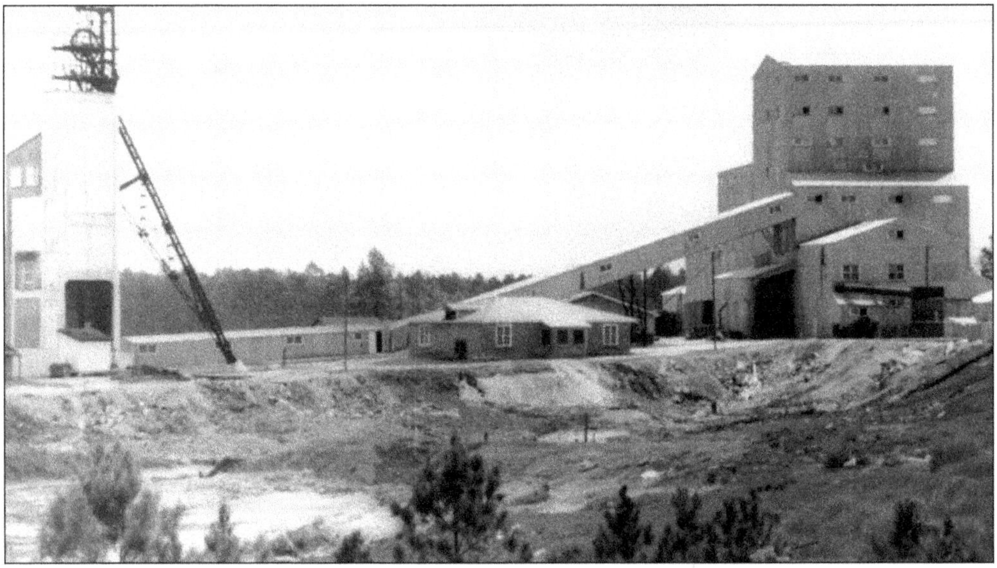

Throughout the 1960s, the AEC expanded its experiments to the Carey Salt Company facilities at Winnfield, Louisiana, pictured here, and at Lyons. An article in a 1959 issue of *Chemical Week* names the Winnfield mine as the target of a series of "nonnuclear high explosive detonations." The article prompted Howard J. Carey to issue a memorandum to the Winnfield employees, stating that the parties were still negotiating and nothing had yet been signed.

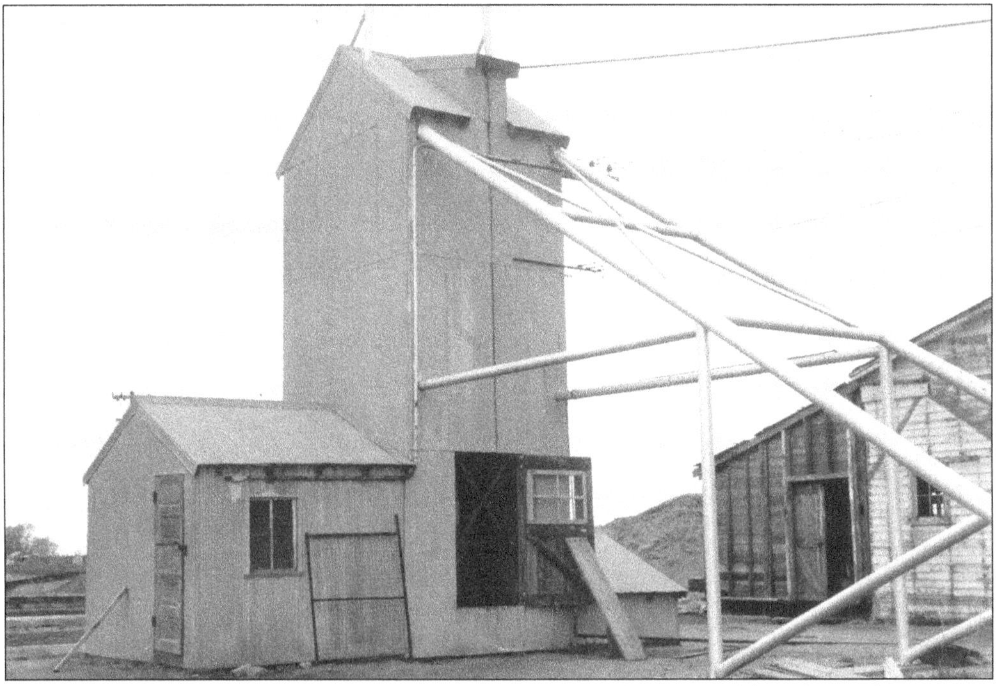

The Lyons plant made a particularly appealing test site because, although employees were keeping it maintained, it was no longer being mined. The AEC decided the scientists would experiment with actual nuclear waste to work out a process for moving and storing the barrels. This image shows the Lyons shaft as it looked in 1959, after the aboveground mill had been dismantled due to lack of activity.

This 1966 photograph shows a container of atomic waste before it is sent into the mine at Lyons. The metal building sat over a specially drilled, 19-inch shaft used to take the canisters into and out of the mine.

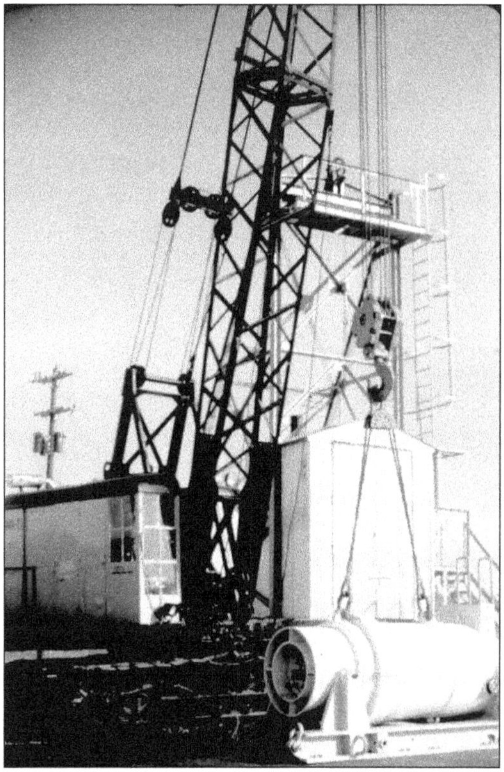

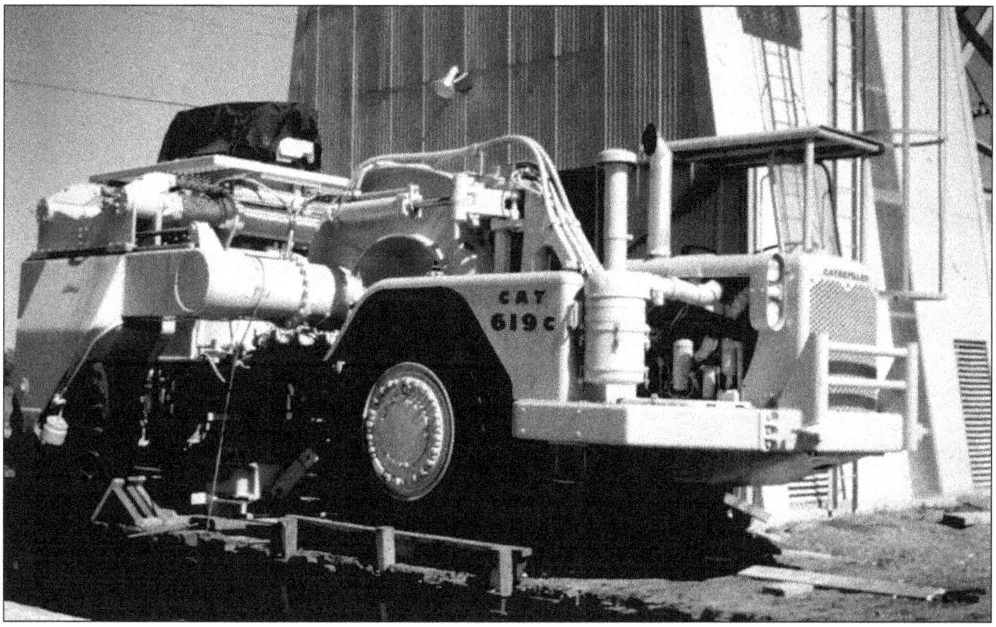

The AEC had a special transporter built to carry the waste to the burial sites underground. Its protective lead shields made it extremely heavy. The largest piece weighed about 11,000 pounds and stretched the hoist cable nine inches when lowered underground in 1964. The transporter cost $100,000 to build, disassemble, and reassemble underground. When the AEC pulled out of Lyons in 1973, it took the transporter with it to New Mexico.

www.ingramcontent.com/pod-product-compliance
Lightning Source LLC
LaVergne TN
LVHW071102050926
847404LV00038B/481

* 9 7 8 0 7 3 8 5 6 1 8 0 6 *